Algal Biology:
a physiological approach

BASIC MICROBIOLOGY

EDITOR: J. F. WILKINSON

VOLUME 9

Algal Biology: a physiological approach

W. MARSHALL DARLEY
Department of Botany
The University of Georgia

BLACKWELL SCIENTIFIC PUBLICATIONS

OXFORD LONDON

EDINBURGH BOSTON MELBOURNE

© 1982 by
Blackwell Scientific Publications
Editorial offices:
Osney Mead, Oxford, OX2 0EL
8 John Street, London, WC1N 2ES
9 Forrest Road, Edinburgh, EH1 2QH
52 Beacon Street, Boston
 Massachusetts 02108, USA
99 Barry Street, Carlton
 Victoria 3053, Australia

All rights reserved. No part of this publication may be reproduced, stored in a retrieval system, or transmitted, in any form or by any means, electronic, mechanical, photocopying, recording or otherwise without the prior permission of the copyright owner

First published 1982

Set by Santype International Ltd, Salisbury and printed and bound in Great Britain by Butler & Tanner Ltd, Frome and London

DISTRIBUTORS

USA
 Blackwell Mosby Book Distributors
 11830 Westline Industrial Drive
 St Louis, Missouri 63141

Canada
 Blackwell Mosby Book Distributors
 120 Melford Drive, Scarborough
 Ontario, M1B 2X4

Australia
 Blackwell Scientific Book Distributors
 214 Berkeley Street, Carlton
 Victoria 3053

British Library Cataloguing in Publication Data
Darley, W. Marshall
 Algal Biology.—(Basic microbiology; v. 9)
 1. Algae
 I. Title
 589.3   QK566
ISBN 0-632-00608-0

Contents

Preface, viii

1 A survey of the algae, 1
   Algal classification, 1
   Algal reproduction, 3
   The classes of algae, 4
   Cyanophyceae, 4
   Chlorophyceae and Charophyceae, 5
   Euglenophyceae, 7
   Pyrrhophyceae, 8
   Xanthophyceae, 9
   Chrysophyceae, 9
   Bacillariophyceae, 10
   Phaeophyceae, 11
   Eustigmatophyceae, 12
   Prymnesiophyceae, 12
   Cryptophyceae, 13
   Rhodophyceae, 13
   References, 14

2 Concepts in algal physiological ecology, 15
   Limiting factors, 15
   Standing stock versus rate functions, 17
   Descriptive versus experimental studies, 18
   Cultures versus natural systems, 19
   References, 20

3 Phytoplankton: environmental factors affecting growth, 21
   The habitat, 23
   Light, 25
   Photosynthesis versus light intensity, 26
   Sun–shade adaptation, 27
   Light-saturated photosynthesis, 30
   Light and metabolism, 31
   Complementary chromatic adaptation, 31
   Nutrients, 32
   Uptake kinetics, 32
   Nutrient-limited growth, 34
   Physiological adaptations to nutrient-limited growth, 36
   Nutrient status of natural phytoplankton, 39
   Nitrogen, 42
   Phosphorus, 42
   Silicon, 43
   Trace elements, 44
   Vitamins, 45
   Temperature, 45

Salinity, 48
Heterotrophic growth, 49
Organic excretion, 51
References, 51

4 Phytoplankton: population dynamics, 53
Diel periodicity, 53
Suspension, 58
Allelopathic interactions, 59
Red tides, 60
Blue-green algae, 63
Anoxygenic photosynthesis, 63
Nitrogen fixation, 63
Gas vacuoles, 65
Blooms, 67
Resting cells, 67
Competition and the paradox of the plankton, 68
Seasonal succession, 70
Evolutionary ecology, 72
References, 72

5 Benthic algae, 74
Periphyton, 77
Epilithic algae, 77
Epiphytic algae, 79
Epipelic algae, 82
Charales, 85
References, 86

6 Seaweeds, 88,
Environmental factors affecting growth, 91
Light, 92
Nutrients, 94
Temperature, 96
Salinity, 97
Desiccation, 98
Environmental control of development, 100
Translocation, 104
Organic exudation, 104
Community interactions, 105
References, 108

7 Symbiotic associations, 111
Algae–Invertebrate associations, 111
Platymonas and *Convoluta*, 113
Chlorella and *Hydra*, 113
Chlorella and *Paramecium*, 114
Zooxanthella and Coelenterates, 115
Chloroplasts in sea slugs, 119
Lichens, 119
Anabaena–Azolla symbiosis, 126
Nostoc–bryophyte symbiosis, 128
References, 130

8 Algae in soil, snow and hot springs, 132
Soil algae, 132
Snow algae, 135
Hot-springs algae, 136
References, 141

9 Algae and mankind, 143
    Troublesome algae, 143
    Effects of pollutants on algae, 145
    Mass cultivation of algae, 147
    Microalgae, 148
    Seaweeds, 149
    References, 150

   Appendixes, 153
1 Detailed characteristics of algal classes, 153
2 Methods in algal physiological ecology, 155
    Biomass (standing crop), 155
    Primary productivity, 157
    Culture techniques, 159
    References, 162

   Index, 164

Preface

In this book I have attempted to provide an introduction to the fascinating and rapidly developing field of algal physiological ecology. The tremendous diversity in structure and reproduction found among the algae often restricts the time which is available in phycology courses to cover ecological topics. I think this book will be a useful supplement to such courses as well as to other courses in aquatic ecology, limnology and marine biology. In conjunction with recent literature it may serve as the text in a graduate-level algal ecology course.

This summary of our present understanding of algal physiological ecology is organized by habitat with an emphasis on experimental rather than descriptive research. Research from controlled laboratory systems are integrated with results from field studies to help explain the alga's success in the real world. The emphasis is on the individual organism and how it functions in its environment. Current methodology is also reviewed.

I would like to express my appreciation to numerous colleagues and students, past and present, in the Department of Botany, University of Georgia, for their helpful comments and stimulating conversation. I feel a special debt to F. Gerald Plumley for his insight and constructive criticism, to Andrew J. Lampkin III for his drawings of representative algae, to typists Carla Ingram and Kim Potts for their skill and patience and to Priscilla Darley for her moral support and help with the proof-reading. Finally I am indebted to the authors of many excellent publications which provided useful ideas and information but which could not be cited due to space limitations. The works which are cited are generally recent publications by leaders in the field and therefore provide a useful point of entry into the literature.

Athens, Georgia   W. Marshall Darley

1 A survey of the algae

The assemblage of primitive plants which we collectively refer to as algae includes a tremendously diverse array of organisms. Algae vary in size from single cells as small as one micrometre to large seaweeds which may grow to over 50 metres. Most algae are photoautotrophs although some, obviously closely related forms, are facultative or obligate heterotrophs; a few algae are even phagotrophic. While most algae are aquatic in either fresh or marine waters, some species may be found growing in such diverse habitats as tree trunks, snow banks, hot springs or even within minute cavities in desert rocks. A number of algae live symbiotically with (or within) animals, fungi or other plants. Perhaps the most striking indication of the diversity of the algae is that they are placed in three separate kingdoms in the recent five kingdom concept of classification of organisms (Whittaker 1969). Thus it is difficult, at best, to provide an accurate, all-inclusive definition of the algae. The following technical definition of algae given by Bold and Wynne (1978) is based on the characteristics of sexual reproduction: sex organs, if present (i.e. some unicellular algae may function directly as gametes), are unicellular or, if multicellular, every cell is fertile. Even this definition leaves us with difficulties. It certainly must be interpreted liberally when considering the sexual organs of the stoneworts. Furthermore, sexual reproduction is unknown in some groups (e.g. euglenoids).

ALGAL CLASSIFICATION

It is perhaps not very surprising that the classification and evolutionary relationships of such a diverse group of organisms have been and continue to be a subject of much debate. The earliest attempts to distinguish separate groups of algae relied on pigmentation, the importance of which we still recognize when we refer to groups such as the green or brown algae. Today we also consider the following characteristics important in defining major algal groups: presence or absence of flagella; flagellar characteristics (number, length, point of insertion, presence or absence of hairs or scales); cell-wall composition; and type of stored photosynthetic product. As more and more algae are examined with the electron microscope, it is becoming increasingly evident that ultrastructural features, especially details of chloroplast structure, are proving to be very useful taxonomic characters. Using

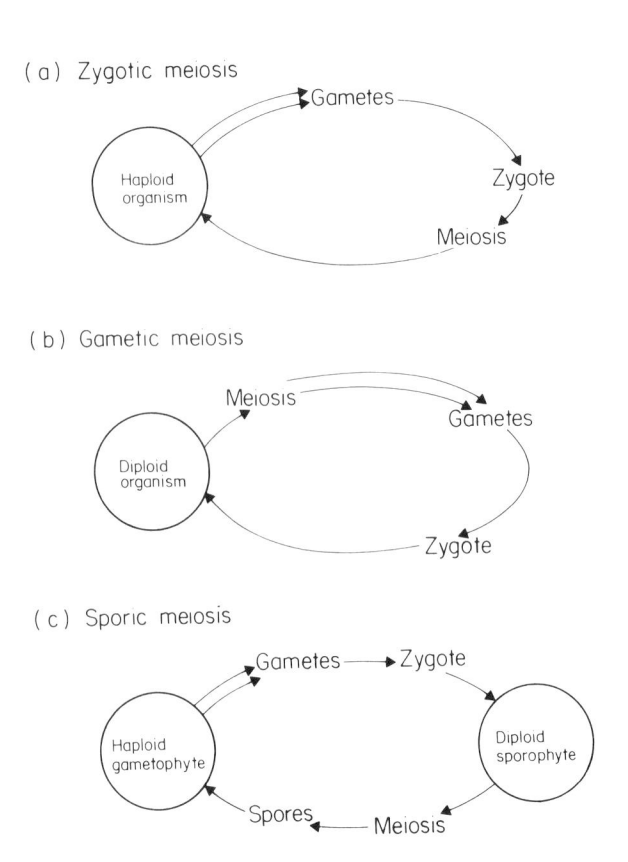

Fig. 1.1 Three, basic, sexual life cycles found in algae. The circle indicates a free-living individual. (After Bold & Wynne 1978.)

of course, many variations on these three, basic, life-cycle patterns, especially among the red algae in which a second diploid phase is often interpolated between the zygote and the sporophyte. The sexual cycles of many algae are as yet poorly known.

THE CLASSES OF ALGAE

The following pages provide an overview of the more salient features of the various classes of algae. The information is meant to complement the details given in Appendix 1 and to provide the student with a brief introduction to the diversity of the algae. For more information the reader should refer to textbooks on algae (Bold & Wynne 1978; Lee 1980; Trainor 1978) or to monographs devoted to certain groups of algae (Carr & Whitton 1973; Cox 1980; Fogg *et al.* 1973; Leedale 1967; Pickett-Heaps 1975; Werner 1977).

Cyanophyceae

The blue-green algae are the only group of algae which possess the prokaryotic type of cellular organization and are accordingly called Cyanobacteria by some. The presence of chlorophyll a and O_2-evolving photosynthesis, however, separates the blue-greens from other photoautotrophic prokaryotes and is cited as justification for continuing to refer to them as algae.

In addition to their widespread occurrence in aquatic habitats, blue-green algae are also common in some extreme environments, especially those that are hot and/or dry. Some even grow within porous desert rocks, in a thin layer near the surface (Friedmann 1971). Blue-greens are found growing symbiotically with fungi (in some lichens) and with several plants. In an ecological context, the blue-green algae are best known for their ability to fix nitrogen, a process discussed in more detail in Chapter 4. Several blue-green algae are known to produce highly toxic endotoxins which are of some economic and ecological importance when these species produce blooms.

Diversity of body form in blue-green algae is limited to unicells, colonies and branched and unbranched filaments. Since morphological characteristics are difficult to discern in these small cells and are known to vary widely with environmental conditions, it has been suggested that physiological and biochemical characteristics be used in classification as they are in bacteria. Sexual reproduction in the traditional sense of fusion of gametes is not known to occur in the blue-greens, although genetic recombination has been reported. Many blue-green algae survive periods of unfavourable conditions through the production of specialized, large, thick-walled, darkly pigmented cells (akinetes) which are full of food reserves. Although blue-greens lack flagella, many exhibit a gliding motility which is thought to somehow involve contractile waves in one of the cell-wall layers. The distinctive blue-green colour is due to the presence of phycocyanin, one of the accessory pigments known as phycobiliproteins. Phycoerythrin may predominate in some blue-green algae to the extent that they appear purple or red rather than blue-green.

Chlorophyceae and Charophyceae

The green algae (considering both classes for the moment) are one of the larger groups of algae in terms of numbers of species and are almost as widely distributed and well adapted to extreme habitats as are the blue-green algae. The green algae have also evolved the greatest diversity of body form found among the algae. With the exception of the stoneworts (e.g. *Chara*), the larger body types tend to be marine and are often tropical with either gametic or sporic meiosis life cycles. The species with smaller body forms are more commonly freshwater or terrestrial with zygotic meiosis life cycles, although several motile unicells are members of the marine phytoplankton.

Both motile and non-motile unicellular and colonial forms are represented among the green algae. Unbranched and branched filamentous forms are common; some species have a well developed, or even dominant, basal, prostrate, pseudoparenchymatous system of filaments from which the erect filaments develop. The familiar sea lettuce (*Ulva*) consists of a blade, two cell layers thick, which is held to the substrate by rhizoidal extensions from basal cells. The forms with siphonous construction consist of long, multinucleate tubes lacking cross walls. Most of these species are macroscopic and found in tropical marine waters; in some species the siphonous filaments are interwoven to produce large plant bodies up to 10–20 cm in length (e.g. *Codium, Halimeda*). In addition to the distinctive siphonous forms, three other orders should be given special notice because they are common and possess unique characteristics. The Oedogoniales (e.g. *Oedogonium* and two related genera) possess a unique form of vegetative cell division in which a circular weakening of the cell wall appears and ruptures with a sudden extension of the cell. The Oedogoniales also have an unusual motile cell with a subapical whorl of flagella. The Cladophorales (e.g. *Cladophora*) are filamentous forms with large multinucleate cells. The Zygnematales (e.g. *Spirogyra* and the desmids) lack flagellated cells but produce amoeboid gametes which fuse in a conjugation process reminiscent of the fungal Mucorales (e.g. the bread mould, *Rhizopus*).

Until recently the Charophyceae have only included the stoneworts and have been considered definite but distant relatives of the rest of the green algae. Recent ultrastructural work on comparative aspects of cell division and zoospore anatomy have resulted in a dramatic revision of evolutionary concepts within the green algae (Pickett-Heaps 1975; Stewart & Mattox 1975). The Charophyceae have been expanded to include green algae in which the teleophase nuclei are held apart by a persistent mitotic spindle. Cytokinesis is accomplished by furrowing (cleavage) in primitive forms (*Klebsormidium*) or by a cell-plate phragmoplast system in more advanced forms (*Coleochaete* and the Charales); presumed intermediate stages are found in *Spirogyra*. Thus the apparent evolution of the higher plant type of cytokinesis (cell-plate phragmoplast system) within this line of green algae suggests that they are closely related to the line which gave rise to land plants (Fig. 1.2).

By contrast, in the majority of green algae (Chlorophyceae) the telophase spindle collapses and the daughter nuclei approach one another. A new set of microtubules (the phycoplast) develops between the nuclei in a plane *perpendicular* to the axis of the now missing mitotic spindle. The phycoplast apparently serves as a guide for cytokinesis between the two daughter nuclei; cytokinesis may occur by furrowing or by cell-plate formation. The Ulvales represent a presumed modification in which precocious cleavage during mitosis eliminates the need for the phycoplast following telophase. Intracellular microtubular arrangements and the flagellar insertion in motile cells also differ subtly in the two groups, resulting in slightly asymmetric motile cells in the Charophyceae and symmetrical motile cells in the Chlorophyceae. As a result of this revision, *Chlamydomonas* is now viewed as a derived form which gave rise to the Chlorophyceae rather than as the putative ancestor of all green algae. It is currently the view that the extant

```
Vascular plants
    ↑
Bryophytes
    ↖  ↖
CHAROPHYCEAE                    CHLOROPHYCEAE

 Charales              Ulvales           Chaetophorales
     ↖                    ↑               and Ulotricales
  Coleochaete                              ↗ Oedogoniales
     ↖
  Zygnematales                            ↗ Chlorococcales
     ↖
  Klebsormidium ↖              → Volvocales

         Pedinomonas ←
```

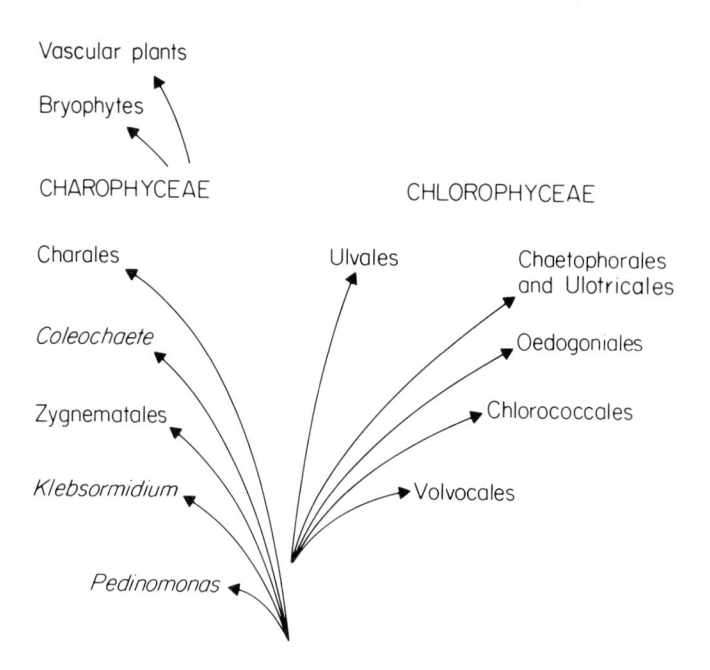

Fig. 1.2 Evolution within the green algae. *Pedinomonas* is a small flagellate; *Klebsormidium* is a uniseriate filament; *Coleochaete* is a disc-shaped epiphyte. Recent evidence on motile-cell ultrastructure suggests that the Cladophorales and Siphonales should be grouped with the Ulvales in a third class, the Ulvaphyceae.

organisms most closely related to the ancestor of the Charophyceae (as well as the Chlorophyceae) will be found among a loosely defined group of motile, unicellular green algae called prasinophytes. Recent research indicates that this is not a natural group; the various species are being placed in one of the two classes of green algae as they are examined ultrastructurally. This very brief account of recent developments in our concepts of evolution in the green algae does not do justice to the insight and the high quality and quantity of work being done by relatively few investigators.

Euglenophyceae

The euglenoids are mostly motile unicells which are not uncommon in both freshwater and brackish habitats. Although there are a fairly large number of different cell types and shapes, there has been no evolutionary development beyond the unicell. One genus (*Colacium*) will develop into a dendroid-shaped colony of non-motile, stalked unicells. Some species which lack emergent flagella retain a degree of motility through the flowing, contracting and expanding euglenoid movement which is characteristic of many of the motile members of the group.

As a group, euglenoids are perhaps the most animal like of the algae. Many species lack chloroplasts and are obligate heterotrophs, either osmotrophic or phagotrophic. The photosynthetic species which have been studied all require at least one vitamin and will supplement their autotrophic nutrition with simple organic acids or alcohols. Many species are incapable of utilizing NO_3^- as a nitrogen source. Thus it is not surprising that euglenoids are most often encountered in environments rich in decaying organic matter (sediment surfaces and in small ponds) and less often found in the phytoplankton of larger bodies of water. It should be stressed that in spite of the vast amount of research which has been carried out on *Euglena gracilis*, few other euglenoid species have been studied in axenic culture. Sexual reproduction in the group is unknown.

Pyrrhophyceae

The dinoflagellates as a group are also rather animal like in a number of respects: most species which have been cultured require one or more vitamins for growth; many heterotrophic forms are known, both free living and parasitic; phagotrophy is not uncommon in both heterotrophic and autotrophic species. Most dinoflagellates are motile unicells common in the phytoplankton of fresh, brackish and marine waters. The few known coccoid and filamentous forms reproduce asexually with the release of the typical dinoflagellate, motile cell characterized by a highly specialized flagellum that usually encircles the cell in a shallow, transverse groove, in addition to the second posteriorly directed flagellum. The chromosomes in the characteristically large nucleus remain condensed throughout the cell cycle giving the interphase nucleus a granular or beaded appearance. Basic proteins are associated with dinoflagellate DNA although the histone proteins characteristic of eukarytoic DNA are missing. During nuclear division the nucleus is penetrated by membrane-lined channels containing microtubules which appear to act as a scaffolding over which the chromosomes, which are attached to the intact nuclear membrane, are moved. Nuclear division in dinoflagellates is, therefore, very unusual, and is assuming importance in our concepts of the evolution of mitosis in eukaryotes.

A few dinoflagellates contain a second, typical eukaryotic nucleus, in addition to the characteristic dinoflagellate nucleus. Careful ultrastructural studies have shown that the eukaryotic nucleus is part of a membrane-bound endosymbiont within the dinoflagellate cytoplasm. In addition to the nucleus, the endosymbiont cytoplasm contains chloroplasts, mitochondria and ribosomes. The ultrastructural evidence points to a fucoxanthin-containing alga as the endosymbiont, thus explaining the occurrence of fucoxanthin and the absence of peridinin in some dinoflagellates. In fucoxanthin-containing dinoflagellates which lack a typical eukaryotic nucleus, the endosymbiont has presumably been reduced to just the chloroplast. This discovery suggests that some algal groups might have obtained their chloroplasts from another eukaryotic alga, rather than having them evolve from symbiotic prokaryotes.

In addition to other characteristics which make dinoflagellates unique and seemingly isolated from other algal groups, certain marine dinoflagellates are unique among the algae by virtue of their capacity for bioluminescence. The ecological significance of the production of tiny flashes of blue light is not known, but the intensity of bioluminescence is controlled by a circadian rhythm. *Gonyaulax polyedra* is a favourite organism for the study of biological rhythms since it exhibits several other rhythms in addition to the bioluminescence. Dinoflagellates are part of several topics of ecological interest which are considered later in this book including coastal algal blooms called red tides and symbiotic associations with various marine invertebrates. Some marine species produce potent toxins which accumulate in shellfish and may cause paralytic, shellfish poisoning in humans who eat contaminated shellfish.

Xanthophyceae

The next four classes listed in Appendix 1 (Xanthophyceae, Chysophyceae, Bacillariophyceae, Phaeophyceae) share enough common traits to be placed in a single division. The first three groups are commonly grouped in the division Chrysophyta, but the Phaeophyceae are seldom included, presumably because of their larger size. The Xanthophyceae (yellow-green algae) show a high degree of parallel evolution of body form with the green algae and, in fact, were classified as green algae until the early 1900's. The two groups may be distinguished by a negative iodine test for starch in the Xanthophyceae. The absence of starch in the xanthophyte chloroplast gives it a uniform, glassy-green or yellow-green colour which contrasts with the granular appearance of the starch-containing chloroplasts in the green algae. Most xanthophytes are non-motile unicells (coccoid forms), but motile unicells, filaments and siphonaceous forms are also known. Most forms are freshwater or terrestrial on damp soil. *Vaucheria*, a widespread, siphonaceous genus that forms luxuriant, felt-like masses on mud, is often found in estuarine, intertidal habitats.

Chrysophyceae

The chrysophytes are golden brown in colour and are mostly motile unicells or colonies; a few species exhibit amoeboid, coccoid or filamentous body forms. As a group they seem to prefer cold, freshwater habitats. Many species form a distinctive asexual (or rarely sexual) resting stage or cyst called a statospore. This siliceous, flask-shaped structure forms within the cytoplasm after which most of the protoplast migrates inside and the narrow mouth of the flask is sealed with a plug.

The body covering of the vegetative stage varies considerably within the class. *Ochromonas* is a naked, motile unicell which has seen some duty as a laboratory organism for nutritional studies since it supplements its photoautotrophic nutrition with osmotrophy and phagotrophy. The cells of the beautiful, colonial genus, *Synura*, are covered with minute, siliceous scales acid, a uronic acid gel which is used as a filler or stabilizer in a number of products.

Eustigmatophyceae

This very small group of marine and freshwater algae has recently been removed from the Xanthophyceae on the basis of ultrastructural features of the zoospore. The group is named for the unusually large eyespot found in the anterior region of the elongate zoospore. All the species which are presently included in the group are coccoid unicells, most of which reproduce by means of zoospores. The Eustigmatophyceae is considered to be quite distinct from the Xanthophyceae and its relatives and not closely related to any of the other algal groups.

Prymnesiophyceae

The Prymnesiophyceae has also recently been removed from another algal class, the Chrysophyceae. Their most characteristic features include a cell covering of scales and a flagellum-like organelle called the haptonema. The latter structure is the basis of an earlier name for the group, the Haptophyceae. The haptonema contains six or seven microtubules surrounded by three membranes; its function is unknown although it has been suggested to be an attachment organelle. It arises between the two, regular flagella where it may be extended or tightly coiled against the cell and difficult to see. The organic scales which cover the cell are of several types: (1) relatively simple, flat, oval plates; (2) elaborate, basket-shaped scales; and (3) oval plates covered with a layer of $CaCO_3$ crystals, especially on the rim. The calcified scales are termed coccoliths and the cells bearing them are called coccolithophorids. All of these homologous scales are produced in the Golgi body and later extruded to the cell exterior where they remain attached to the plasma membrane. In at least one species of coccolithophorid, the coccolith is one-third the diameter of the cell. The production of such a large structure in a Golgi body and its 'birth' to the cell exterior obviously must involve some distortion of cell structures.

The evolution of body form in this group is limited primarily to motile and coccoid forms, although loosely organized filamentous forms are also known. It is possible that a number of the non-motile benthic forms will ultimately be found to represent alternate phases in the life cycle of motile species. As a group the Prymnesiophyceae favour warmer, marine waters.

Cryptophyceae

The cryptomonads are a small group of motile unicells with a number of unique characteristics and combinations of characteristics which make their origins very obscure. The flagella arise within a furrow which opens near the anterior end of these flattened cells. Although they have a proteinaceous cell covering in common with the euglenoids, the structure of this 'periplast' does not resemble the pellicle of the euglenoids. In spite of the relatively small number of species, the cryptomonads are not uncommon in both freshwater and marine habitats.

Rhodophyceae

Most of the red algae are macroscopic seaweeds which share the rocky, coastal waters with green and brown algae. They are usually smaller and more delicate than the brown algae. Red algae tend to be more abundant in subtropical and tropical waters than in temperate and polar seas and they are often more abundant in deeper waters than the green and brown algae which generally prefer colder water and the intertidal zone. In addition to the more conspicuous fleshy or membranous forms, red algae also occur as unicells, small-branched filaments, crustose forms, and as small parasites on other red algae. The cell walls of a number of red algae are heavily impregnated with $CaCO_3$. Although we think of red algae as seaweeds, they are also fairly common in fresh water; however they tend to be restricted to running water. Simply because red algae are placed last in many phycology courses and text books, one must not assume that red algae are the most highly evolved of the algae. In spite of their size and extremely complex life cycles, red algae have a number of characteristics that may be considered primitive including the absence of flagellated cells, simple chloroplast structure and the presence of phycobilin pigments. In most species the thallus consists of an aggregation of filaments rather than a true parenchyma.

Among phycology students, red algae are infamous for the complexity of their life cycles. In addition to the gametophyte (haploid) and sporophyte (diploid, called a tetrasporophyte in the red algae) generations in the usual alternation of generations type of life cycle (Fig. 1.1c), most of the red algae insert a second, diploid phase following zygote formation and preceding tetrasporophyte development. This post-fertilization development, including various taxonomically significant cell fusions and divisions, results in the growth of a small, diploid plant, the carposporophyte, on the female gametophyte. The carposporophyte generation converts every successful fertilization event into hundreds of diploid cells (carpospores) prior to their release from the gametophyte. Assuming the chances of fertilization are decreased in this algal group by the absence of flagellated cells, the carposporophyte generation represents an effective means of increasing reproductive potential (Searles 1980).

The red algae have been exploited for commercial applications in several ways. A number of species are utilized for food in the Far East and other countries. The membranous *Porphyra* in particular is intensively cultivated in Japan and eaten as 'nori'. Agar, the familiar gelling agent for culture media, is a complex polysaccharide containing galactose and galactose derivatives. Carrageenan is a sulphated galactan which is utilized as a stabilizer and emulsifier in dairy products, toothpastes, cosmetics and a number of other products.

REFERENCES

BOLD H.C. & WYNNE M.J. (1978) *Introduction to the Algae*. Prentice-Hall, Englewood Cliffs, New Jersey.
CARR N.G. & WHITTON B.A. (eds) (1973) *The Biology of Blue-Green Algae*. Blackwell Scientific Publications, Oxford.
COX E.R. (ed.) (1980) *Phytoflagellates*. Elsevier/North-Holland, New York.
FOGG G.E., STEWART W.D.P., FAY P. & WALSBY A.E. (1973) *The Blue-Green Algae*. Academic Press, New York.
FRIEDMANN E.I. (1971) Light and scanning electron microscopy of the endolithic desert algal habitat. *Phycologia*, **10**, 411–28.
GIBBS S.P. (1978) The chloroplasts of *Euglena* may have evolved from symbiotic green algae. *Can. J. Bot.*, **56**, 2883–9.
LEE R.E. (1980) *Phycology*. Cambridge University Press, Cambridge.
LEEDALE G.F. (1967) *Euglenoid Flagellates*. Prentice-Hall, Englewood Cliffs, New Jersey.
LEWIN R.A. (1976) Prochlorophyta as a proposed new division of algae. *Nature (Lond.)*, **261**, 697–8.
PICKETT-HEAPS J.D. (1975) *Green Algae: Structure, Reproduction and Evolution in Selected Genera*. Sinauer Associates, Sunderland, Massachusetts.
RAVEN P.H. (1970) A multiple origin for plastids and mitochondria. *Science*, **169**, 641–6.
SEARLES R.B. (1980) The strategy of the red algal life history. *Am. Nat.*, **115**, 113–20.
STEWART K.D. & MATTOX K.R. (1975) Comparative cytology, evolution and classification of the green algae with some consideration of the origin of other organisms with chlorophylls *a* and *b*. *Bot. Rev.*, **41**, 104–35.
TRAINOR F.R. (1978) *Introductory Phycology*. John Wiley & Sons, New York.
WERNER D. (ed.) (1977) *The Biology of Diatoms*. Blackwell Scientific Publications, Oxford.
WHITTAKER R.H. (1969) New concepts of kingdoms of organisms. *Science*, **163**, 150–60.

2

Concepts in algal physiological ecology

The remaining chapters in this book consider the physiological ecology of algae which grow in a variety of habitats. The focal point will be the individual alga and the ways in which it is physiologically, structurally and reproductively adapted to its environment. This approach is often called autecology. The broader view of algal ecology will not be considered even though it is recognized that algae are often very important components of their ecosystems, making significant contributions to nutrient and carbon cycles and to the flow of energy. The holistic approach to algal ecology is summarized in Volume 5 of this series and in several texts on aquatic ecology (Wetzel 1975; Parsons *et al.* 1977).

Algal species are successful in their habitats by virtue of their own unique genetic programming which allows them to: (1) fulfill their basic physiological requirements for growth and reproduction within a particular environment; (2) successfully interact with other organisms in that environment; and (3) survive periods during which they may not be able to grow due to unfavourable conditions in that environment. The algal ecologist would like to understand these requirements, interactions and adaptations so that the organism's presence and ecological significance in time and space can be explained and predicted. Ultimately, we hope that our understanding of the environmental biology of algae can be integrated with data on the dynamics of other functional components of the ecosystem to yield a more comprehensive understanding of ecosystem structure and function. The remainder of this chapter considers conceptual approaches to the study of algal physiological ecology. Some of the more commonly applied techniques which are currently available for these studies are described in Appendix 2.

LIMITING FACTORS

The limiting factor is the essential resource (out of the entire spectrum of essential resources) that is present in quantities most closely approaching the critical minimum needed by the organism, in this case an alga (Odum 1971). It follows that if the level of the limiting factor were increased, the algal population would grow at a faster rate until the level of some other essential resource limited the growth rate. The concept of limiting factors is most readily applied under steady-state conditions (Odum 1971).

As originally developed by Liebig, the concept of limiting factors only applied to deficiencies of chemical nutrients (the 'Law of the Minimum'). Others have expanded the concept to include physical parameters such as light intensity and temperature. The concept was further broadened by Shelford when he recognized that environmental parameters could limit growth in their excess (high temperatures, high concentrations of NH_4^+) as well as in their deficiency. The maximum and minimum tolerable levels thus define the *tolerance limits* of the organism to any environmental factor. Other organisms can be considered limiting if they produce toxic materials with sufficient potency or in sufficient concentration to inhibit an alga's growth. Grazing organisms can also be limiting in the sense that they continually harvest the algae and thus may prevent the standing crop from increasing. Note the distinction here between limiting the standing crop and limiting the growth rate. Although they lower the standing crop, zooplankters often increase the rate of nutrient recycling and therefore may actually stimulate the growth rates of the surviving algae when nutrient levels are limiting.

It is also important to realize that what is limiting for one species (or ecotype) is not necessarily limiting for another. Therefore, a measure of community activity (e.g. primary productivity) could be limited by more than one resource. Even when considering a single species, it is sometimes useful to distinguish between limitations at different levels of the organism's biology. Is the condition in question limiting the growth rate of the organism or is it preventing any growth at all or is it even more severe and threatening its survival? In other words, how is the balance between population growth and population decline affected by the factor in question? Is the condition limiting for sexual reproduction but not for continued growth and asexual reproduction? The answers to these questions involve short-term as well as long-term considerations.

At first glance it might seem that one could explain and predict an alga's success in nature by defining optimum conditions and limits of tolerance of that alga for all biotic and abiotic environmental parameters. While this approach will provide a great deal of information, it will not provide the complete answer we seek. Describing optimum conditions and tolerance limits for the large number of known parameters would be, at best, a formidable task; there are no doubt some important biotic and/or abiotic factors of which we are not yet even aware. The task is made more difficult by the fact that optimum conditions and/or tolerance limits for one parameter often will change as a second parameter fluctuates. Being successful in a given ecosystem almost always means that the alga must be flexible enough in its requirements and interactions to compromise, enduring a less than optimum (but still within tolerance limits) situation with respect to one parameter in return for a more favourable (and now within tolerance limits) situation with respect to other parameters. Therefore, under most natural situations, the performance of the alga under suboptimum conditions will be of greater ecological significance than its performance under optimum conditions. In addition, an alga's performance may be affected by environmental conditions of the immediately preceding hours or days. For example, its ability to store inorganic nutrients and photosynthetic products during favourable periods may enhance its performance under less favourable conditions at a later time. Finally, algae are often capable of adapting or acclimating so as to improve their performance as conditions change.

The preceding discussion was confined for the most part to the physiology of the alga, although morphological features and aspects of the reproductive cycle (sexual as well as asexual) are also very important to the success of an alga in its habitat. Long-term survival under adverse conditions often depends on the production of resistant resting spores of some type. Clearly the ecologist is faced with an extraordinarily complex set of inter-relationships when setting out to define and describe the factors which account for an alga's success at any point in space and time. When one considers the number of different species present in most systems, the situation is truly overwhelming. The principal usefulness of the concept of limiting factors to physiological ecology is that it allows us to isolate individual parameters among the broad spectrum of complex inter-relationships and to focus our research efforts on them. Factors which are limiting are most likely to be those which are 'operationally significant' (Odum 1971) as controlling mechanisms, determining the success of the organism in a given environment. Other factors which are at more nearly optimum levels, need not be critically evaluated, since they are less likely to exert controlling influences. We are still left with an impressive list of interactions to investigate, but at least the list has been shortened.

STANDING STOCK VERSUS RATE FUNCTIONS

A second important concept to keep in mind is the distinction between the amount of material present at any one time and its rate functions (standing crop versus rate of exchange or turnover). As an example, suppose that a benefactor offered to supply your groceries under either Plan A (one bag of groceries) or Plan B (ten bags of groceries). An ecologist would want to know how often deliveries were made before making a choice between the two plans (assuming the rate of consumption would be the same in either case). What if deliveries were once a week on both plans? What if deliveries were once a week on Plan A and once a year on Plan B? The point here is that the significance of an ecosystem component is not always determined by standing stock alone. Even though standing crops are usually much easier to measure than rate functions, a complete understanding of the system often requires some information about the rate of supply. This concept is especially important when dealing with micro-organisms. Since metabolic activity per unit biomass increases as the cell (or organism) size decreases, microbes may contribute much more to community processes than their biomass would indicate. Thus bacteria in a pond may be a small fraction of total community biomass and yet be responsible for a large fraction of total community respiration. It is also important to remember rate functions when considering nutrients as limiting factors. Even though the rate of nutrient uptake at limiting concentrations will depend on the nutrient standing stock, the level of that standing stock is a function of: (1) the rate of supply through mineralization and other processes; and (2) the rate of assimilation by the total community.

DESCRIPTIVE VERSUS EXPERIMENTAL STUDIES

In order to help organize the discussion of the autecology of algae, it is useful, at the risk of oversimplification, to classify the research approach as either descriptive or experimental. In a descriptive study, the aim is to describe the system as it exists in its natural state. ('Let's take a peek and see who's there and what's happening.') This approach, which is most often applied to natural systems, should provide a description of both standing state and dynamic features of the system with a minimum disturbance to its natural status. At the organism level, the factors controlling the success of the alga in question are determined through correlations between standing state and rate functions of the alga and various prevailing environmental parameters, including the possible influence of other organisms.

In the experimental study, the system is manipulated in some way and the effects of the manipulation are observed. ('Let's turn off the lights and see what happens.') This type of study may be done with laboratory cultures or with natural systems and often tests hypotheses generated in descriptive studies. The experimental manipulation should involve suspected limiting factors, including biotic components such as the number of algal cells or the abundance of zooplankton grazers, as well as physical and chemical environmental parameters such as temperature, light intensity or nutrient concentrations. The effects may be monitored as standing state and/or dynamic aspects of the system, preferably both. Thus the experimental approach provides more direct evidence for the various physical, chemical and biological factors which are responsible for the success of the alga in natural systems. The closer the experimental system and the manipulations mimic natural systems, the more meaningful the results. Since it is often difficult to manipulate morphological features and the sexual life cycle in algae, the survival benefit of these characteristics is often deduced from descriptive studies alone.

Both the descriptive and experimental approaches are essential to a complete understanding of algal ecology. In both approaches the research design and interpretation of results often focus on a consideration of limiting factors. In a descriptive study, one takes advantage of naturally occurring gradients of environmental parameters either in space or time. Regression analysis identifies parameters which are highly correlated with algal performance; these parameters are likely candidates for limiting factors. In experimental studies the investigator provides the gradient. If a minor adjustment in a given environmental parameter results in a significant change in the alga's performance, that parameter is considered limiting. Minor adjustment of non-limiting parameters will have little, if any, effect on the alga. The advantage of the experimental system is that the investigator can manipulate one factor of his choosing at a time, whereas in a descriptive study on a natural system, several parameters are likely to be changing simultaneously, making it difficult to ascribe effects to any particular environmental factor. A disadvantage of the experimental system is that by manipulating the system, it is no longer in its 'natural' state.

CULTURES VERSUS NATURAL SYSTEMS

Since research on algal autecology utilizes both natural and culture systems, one should appreciate the advantages and limitations of each. The principal advantage of a natural system is that it represents the actual set of conditions under which the alga is or is not successful, and that, after all, is what we are trying to understand. In a sense, a natural system represents the ultimate definition of a suitable environment for those organisms which are successful in it. The fact that we cannot put this definition on paper is one of the major limitations of the use of natural systems. It is also difficult to reproduce natural systems for comparative studies, especially if one is trying to repeat an experiment at a later date. Due to the variability or 'patchiness' in natural systems, many replicate samples must often be taken to obtain a reasonably accurate estimation of the mean value of the parameter being measured. When comparing the mean of one data set with that of another, statistically derived confidence limits are used to determine the probability that the true means (assuming a very large number of samples have been taken) of the two data sets are different. For example, suppose the photosynthetic rate in ten replicate samples to which nutrients had been added was measured and that data set compared with photosynthetic rates in ten samples from the same water mass but without added nutrients. The mean of the samples with nutrients is found to be higher than the mean of samples without nutrients, but it is unclear whether the difference is significant because of the high variability within each data set. Statistical analysis might show that in 95% of such comparisons the means would not be identical. In such a case the nutrient-enriched samples would have 'significantly' higher rates. However, if the value were 99%, the difference in data sets would be 'highly significant.'

Cultures of algae are useful principally because the alga can be studied under defined and reproducible conditions. Culture studies have contributed tremendously to our understanding of algal life cycles, nutrition, chemical composition and growth rates. Results are often less ambiguous than when working with natural samples, but statistical analysis should also be employed. The principal limitation of cultures is that they do not represent natural conditions and are, therefore, 'unnatural'. The culture is either unialgal (bacteria present) or axenic (a pure culture) and thus the natural interactions with other organisms are reduced or absent. Nutrient levels are usually 100–1000 times higher than in nature and thus older cultures suffer extreme and 'unnatural' crowding. Species which grow well in culture (sometimes referred to as laboratory weeds) are not always the same species which are most abundant or 'important' in natural systems. These problems have caused some workers to question the use of cultures in ecological research. A compromise between the need for defined conditions and the problem of unnatural conditions may be reached by more closely simulating the natural environment in the culture vessel. Continuous culture and dialysis culture systems are major improvements in this regard and are discussed in Appendix 2. In conclusion we must remember that there are significant ecological differences between cultures and natural systems and that we should be careful in our comparisons and extrapolations from one to the other, but that this cautious approach must not prevent us from taking advantage of the benefits offered by each. An attempt is made in this book to consider and compare results from both types of studies.

REFERENCES

ODUM E.P. (1971) *Fundamentals of Ecology*, 3rd edn. W.B. Saunders Co., Philadelphia.
PARSONS T.R., TAKAHASHI M. & HARGRAVE B. (1977) *Biological Oceanographic Processes*, 2nd edn. Pergamon Press, Oxford.
WETZEL R.G. (1975) *Limnology*. W.B. Saunders Co., Philadelphia.

3 Phytoplankton: environmental factors affecting growth

Planktonic organisms are suspended in the water column and lack the means to maintain their position against the current flow, although many of them are capable of limited, local movement within the water mass. Phytoplankton (autotrophic plankton) consists primarily of microscopic, unicellular or colonial algae (Fig. 3.1). They are grazed mainly by zooplankton, a group of organisms which is dominated by rotifers and crustaceans, especially cladocerans and copepods. Zooplankton also includes representatives of many of the other major animal phyla.

The most obvious algae in the plankton are those larger forms which are captured and concentrated in tows with standard plankton nets (pore size of approximately 50 μm). The most abundant of these 'net plankton' in both fresh and marine waters are diatoms and dinoflagellates; centric diatoms outnumber pennate species in the sea, whereas both are common in lakes. The net plankton in fresh waters will also commonly include colonial green algae, desmids, filamentous and colonial blue-green algae and colonial chrysophytes. A filamentous blue-green alga is sometimes found in tropical seas. Organisms which pass through clean plankton nets and are therefore smaller than 50 μm are termed nanoplankton. The distinction is an arbitrary one based on cell size and not all workers use 50 μm as the dividing point. The nanoplankton consists of unicellular blue-green algae and small, often motile unicells belonging to the Chlorophyceae, Chrysophyceae, Cryptophyceae, and Prymnesiophyceae. Small diatoms and dinoflagellates are also included in the nanoplankton.

It is not surprising that the net plankton have received more attention, both in the field and in culture than the nanoplankton species, which in addition to their small size, are very fragile and are often unrecognizable in preserved samples. Our relative ignorance of the nanoplankton is especially serious since evidence continues to accumulate that they are often responsible for the major portion of planktonic primary production, particularly in the sea. Typical cell concentrations of total phytoplankton range from less than 10 cells ml^{-1} in oligotrophic (nutrient-poor) waters to more than 10 000 cells ml^{-1} during spring blooms and over 50 000 cells ml^{-1} in red tides. Cell concentrations in cultures typically range from 10^4 to 10^7 cells ml^{-1}.

Perhaps a brief explanation is in order concerning the relatively large

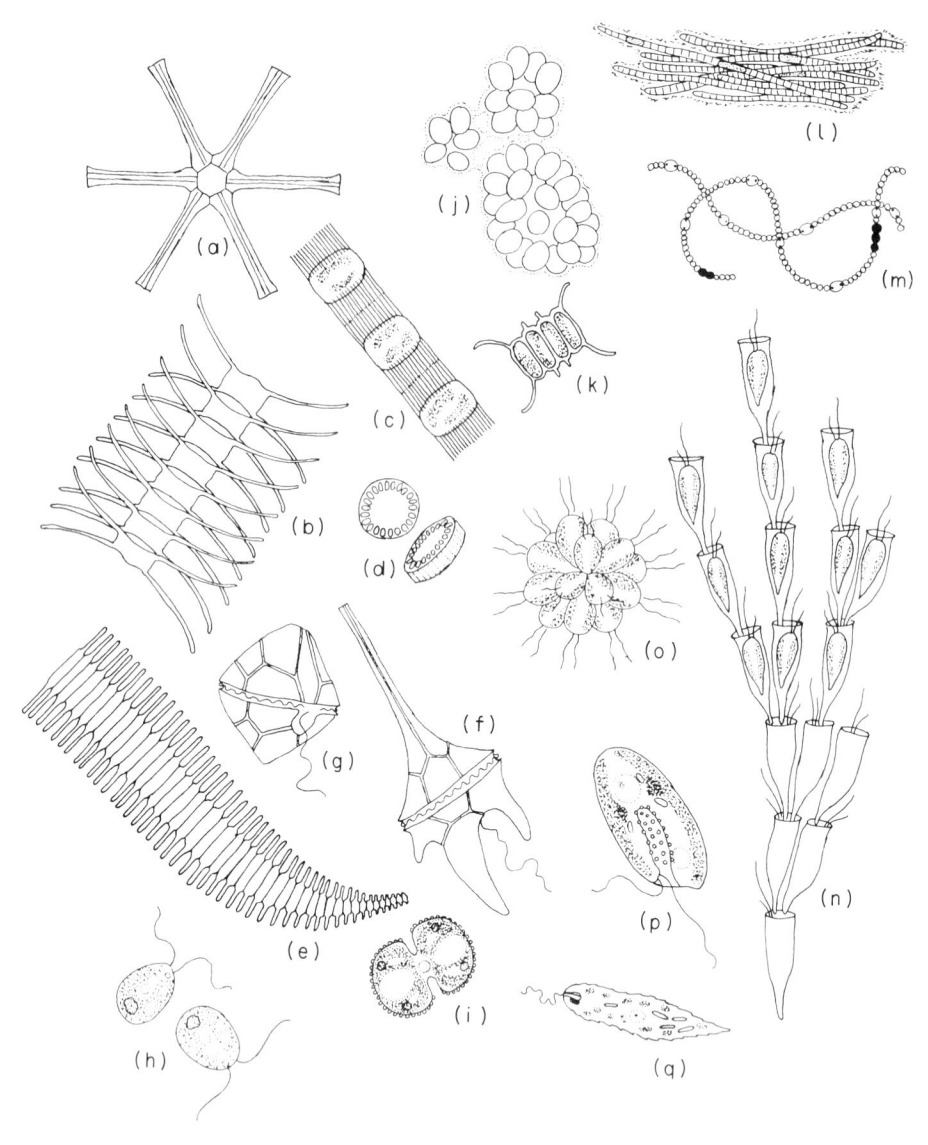

Fig. 3.1 Examples of planktonic algae. Diatoms: (a) *Asterionella*; (b) *Chaetoceros*; (c) *Skeletonema*; (d) *Cyclotella*; and (e) *Fragilaria*. Dinoflagellates: (f) *Ceratium* and (g) *Peridinium*. Green algae: (h) *Chlamydomonas*; (i) *Cosmarium*; (j) *Botryococcus* and (k) *Scenedesmus*. Blue-green algae: (l) *Aphanizomenon* and (m) *Anabaena*. Chrysophytes: (n) *Dinobryon* and (o) *Synura*. Cryptomonads: (p) *Cryptomonas*. Euglenoids: (q) *Euglena*. (Drawings by Andrew J. Lampkin III.)

amount of space devoted to phytoplankton in this book. The principal reason is that more is known about the physiology of planktonic algae than is known about the physiology of algae from any other habitat. Many of these microscopic algae have been grown and studied in suspension culture for a number of years, using techniques developed earlier by microbiologists. Phytoplankton populations in nature are relatively easy to study in that they resemble very dilute cultures and so it is relatively easy to obtain replicate samples; less extraneous living and non-living material is present than in algal samples from other habitats; and the environment is relatively homogeneous and therefore easier to define and reproduce than that in other algal habitats. No doubt the fact that phytoplankton production is estimated to account for over half of the total world's primary production also has provided a significant impetus to the study of phytoplankton physiology. The recent volume edited by Morris (1980) is recommended to those readers wishing to explore the physiological ecology of phytoplankton beyond the present treatment.

THE HABITAT

Phytoplankton must fulfil their ecological requirements for growth within the illuminated (euphotic) surface waters of lakes and seas. The habitat is relatively stable and homogeneous (or unstructured) compared with other algal habitats, although gradients of important environmental parameters do exist in time and space. Since mixing of surface waters is largely a function of wind action, the water column will tend to exhibit more structure and steeper gradients during periods of calm weather. The relative depths of the mixed layer and the euphotic zone are of obvious importance to suspended photoautotrophic cells which must receive light energy to grow. Mixing also serves to replenish nutrients in surface waters. The latter function is especially pronounced along the western shores of some continents where winds and the effects of the earth's rotation combine to move surface waters offshore. The nutrient-rich, 'upwelling' water which replaces the surface water contributes to the high productivity of these areas.

Temperature is an extremely important variable in the mixing process because of its well-known effect on the density of water and therefore on the stability of the water column (Fig. 3.2). In temperate lakes and oceans during the winter, temperature (and therefore density) differences between surface and deeper waters usually are not sufficient to prevent complete mixing by the wind. As surface waters are warmed during the spring and early summer, they become less dense and less likely to be mixed with deeper, colder and denser water. Eventually, density differences are great enough to prevent mixing and the water column becomes stratified into an upper mixed layer (called the epilimnion in lakes) which is separated from the bottom water (hypolimnion) by a zone in which the temperature (and density) of the water changes abruptly. The significance of this zone, called the thermocline, is that it severely restricts the mixing of the deeper, nutrient-rich water with the surface, mixed layer where phytoplankton growth has stripped the water of nutrients. Thus two important requirements for algal growth, light and nutrients, are most abundant in separate water masses. Stratification breaks down in the autumn when surface waters cool and wind-induced circulation results in the autumn overturn. Winter stratification may also occur in lakes, particularly if there is a cover of ice.

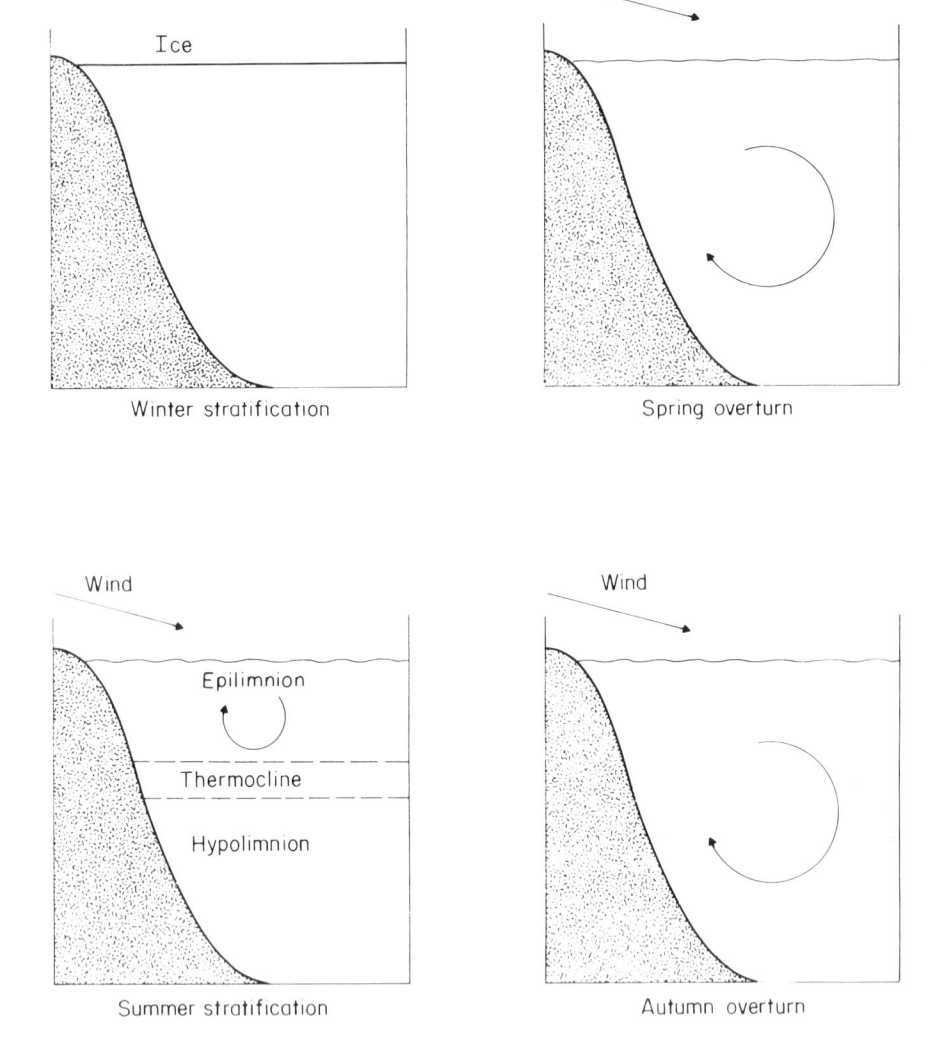

Fig. 3.2 Effect of seasonal temperature changes on vertical mixing in a temperate lake.

This annual cycle of thermal stratification and overturn and the accompanying changes in nutrient concentrations are in part responsible for the spring, and sometimes autumn, phytoplankton blooms in temperate waters. Low temperatures and light intensities limit phytoplankton production in the winter. In polar seas, a thermocline does not develop and productivity is highest during the summer months when light is most intense. In tropical seas, there is a permanent thermocline and the surface waters remain nutrient deficient throughout the year. A variety of stratification patterns are observed in tropical lakes.

Although a number of wind-dependent, mixing processes have been identified in lakes and oceans, the discussion here will be limited to Lang-

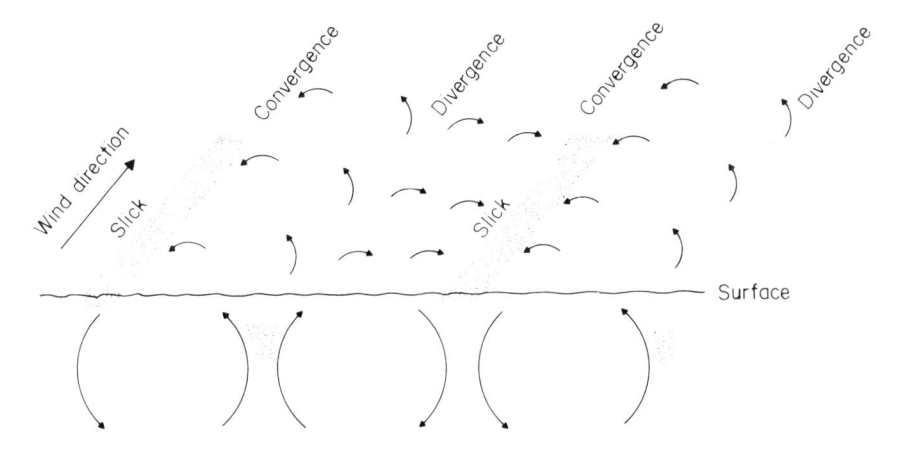

Fig. 3.3 Langmuir circulations.

muir circulations (Fig. 3.3). The action of wind blowing across surface waves sets up a series of rotating convection cells whose axes are parallel to the wind direction. Water moves up or down in areas where adjoining cells diverge or converge, respectively, at the surface. Substances or cells which are positively buoyant (float) tend to accumulate at zones of convergence where they may form visible streaks, windrows or slicks on the water, whereas particles which tend to sink will accumulate at a zone of divergence. Langmuir circulations are primarily responsible for circulation within the mixed layer of lakes and oceans. Convention cells are modified by wind speed and direction and may vary in scale from 1 m to 100 m; cells of different sizes may be present at the same time. Although it is difficult to predict the effect of Langmuir circulations on any given algal cell, it is clear that these currents will have a significant impact on phytoplankton production by carrying cells below the euphotic zone in addition to suspending them within it and by concentrating cells in certain areas where zooplankton grazing could be more efficient.

LIGHT

Light, which must be considered in terms of photoperiod and quality (wavelength) as well as intensity, is of paramount importance to algae. In addition to its obvious significance as an energy source for photosynthesis, the fact that light fluctuates tremendously in both space (depth and latitude) and time (daily and seasonally) suggests that light will often be limiting for phytoplankton growth. Light intensity decreases exponentially with depth in the water column, with blue-green light (480 nm) penetrating the furthest in clear waters. Thus, throughout much of the euphotic zone, accessory pigments such as carotenoids, biliproteins and accessory chlorophylls will assume major roles in absorbing light energy. Day length underwater may be considerably shorter than aboard ship due to the reflection of light from

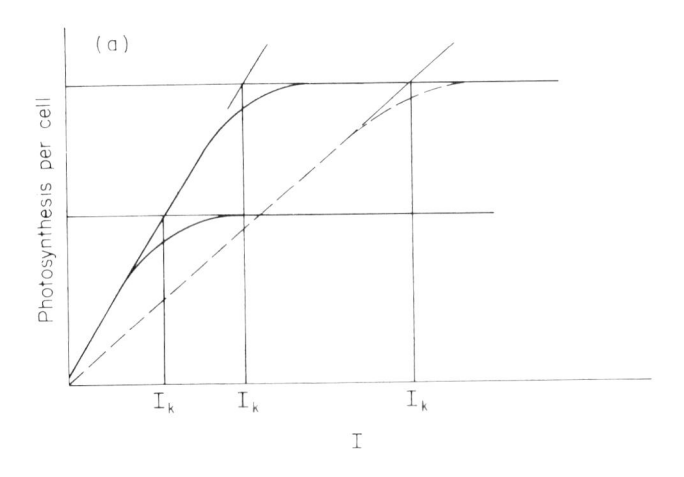

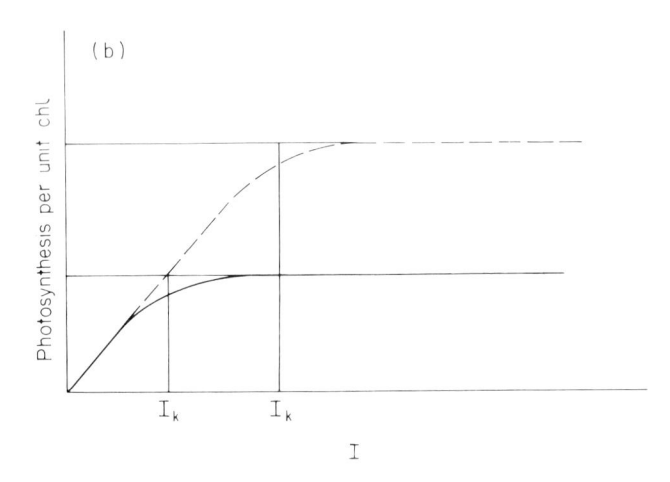

Fig. 3.5 P versus I curves for shade-adapted (solid line) and sun-adapted (dashed line) cells, expressed as photosynthesis per cell (a) and photosynthesis per unit chlorophyll (b).

synthesis at the expense of synthesis of the dark reaction enzymes which contribute to P_{max}. Cells grown in light–dark cycles tend to have shade characteristics as compared with cells grown in continuous light of the same intensity. Although shade-adapted cells are more efficient at absorbing light energy, the quantum yield (mol O_2 einstein^{-1}) remains constant, suggesting they cannot utilize the absorbed light with increased efficiency (Senger & Fleischhacker 1978).

Less is known of how shade adaptation affects the compensation point due to uncertainties and difficulties in measuring photosynthesis and respiration at very low light intensities. The available data suggest that the rate of dark respiration declines during light-limited growth. As a result, the compensation light intensity decreases, an obvious adaptive advantage. The frequent occurrence of large populations of phytoplankton below the 1% light level in nature suggests that at least some species have remarkable capacities to grow at very low light intensities.

At high light intensities plenty of light is available and cell growth is limited instead by the rate at which carbon is fixed, which in turn depends on cell metabolism. Thus, sun cells devote less of their resources to chlorophyll synthesis and more to the synthesis of RuBP carboxylase and possibly other enzymes of the dark reactions. As a result, sun cells are typified by less chlorophyll per cell (sometimes correlated with a decrease in chloroplast size), a lower initial slope, a higher I_k and often a higher P_{max}. Sun-adapted cells are also less sensitive to photoinhibition at high light intensities. Adaptation to shade or sun conditions typically requires one to three days.

Note that when P versus I curves are plotted as the photosynthesis per unit chlorophyll (Fig. 3.5b) rather than on a per cell basis (Fig. 3.5a), the initial slopes for sun and shade cells are superimposed. This result is to be expected if the light reactions are indeed independent of most environmental factors as is generally assumed. Deviations from this expected pattern have been reported, however, in cultures (Falkowski & Owens 1980) and in natural phytoplankton populations (Platt & Jassby 1976). In both these studies the slope decreased with shade adaptation, a change opposite that expected in cells presumably adapted to maximize light absorption. The adaptive significance of these changes in the slope (per unit chlorophyll) are not yet clear, but several mechanisms have been proposed. Changes in chlorophyll a/accessory pigment ratios during sun–shade adaptation would alter the light-capturing efficiency per unit chlorophyll *a* and therefore change the slope. It is also possible that smaller cells will have higher efficiencies than larger cells at identical chlorophyll/carbon ratios because there would be less self-shading of chlorophyll molecules in the small cells. The same effects might be observed in a single species; i.e. as the chlorophyll concentration decreases during sun adaptation, decreased self-shading within the same sized cell could increase the slope.

The shift to shade-cell characteristics may be due as much to the absence of longer wavelength light as to a decrease in light intensity. Vesk and Jeffrey (1977) studied the effect of blue-green light (the predominant wavelengths at the bottom of the euphotic zone) on chloroplast characteristics in 17 species of marine phytoplankton (from six different classes) compared with white light of the same intensity. Six species, mostly diatoms, showed major (up to 146%) increases in chlorophyll per cell in blue-green light while other species showed less pronounced or no changes in chlorophyll content. The large increase in chlorophyll was associated with an increase in thylakoids per chloroplast and an increase in chloroplast number in at least one centric diatom. It seems clear that future laboratory studies on mechanisms of shade adaptation should consider ecologically significant wavelengths as well as intensities of light.

Light-saturated photosynthesis

Phytoplankton in nature may experience a variety of light regimes during the dawn–noon–dusk photoperiod depending on cloud cover, Langmuir circulations (see Fig. 3.3) or even vertical migration. Marra (1978) has investigated one short-term response of P_{max} to these variations in light intensity. The value of P_{max} was found to be a time-dependent function of the light intensity experienced by the cells within the last several hours. (This response should not be confused with the change in P_{max} occurring over one to three days in sun–shade adaptation). Figure 3.6 is a diagrammatic interpretation of Marra's results. Cells which recently have experienced darkness or low light intensities have a relatively high P_{max} which declines after a short period at high light intensities. At lower, but still saturating light intensities, P_{max} also declines but only after a longer period of time. The initial slope of the P versus I curve does not change during these brief time intervals. The adaptive significance of this time dependence for the value of P_{max} appears to be that it allows cells to take advantage of brief periods of high light intensity. With continued exposure to high intensities, however, photosynthetic rates decline to achieve a balance between carbon uptake and overall growth processes.

Possible short-term variations in P_{max} are of great importance in phyto-

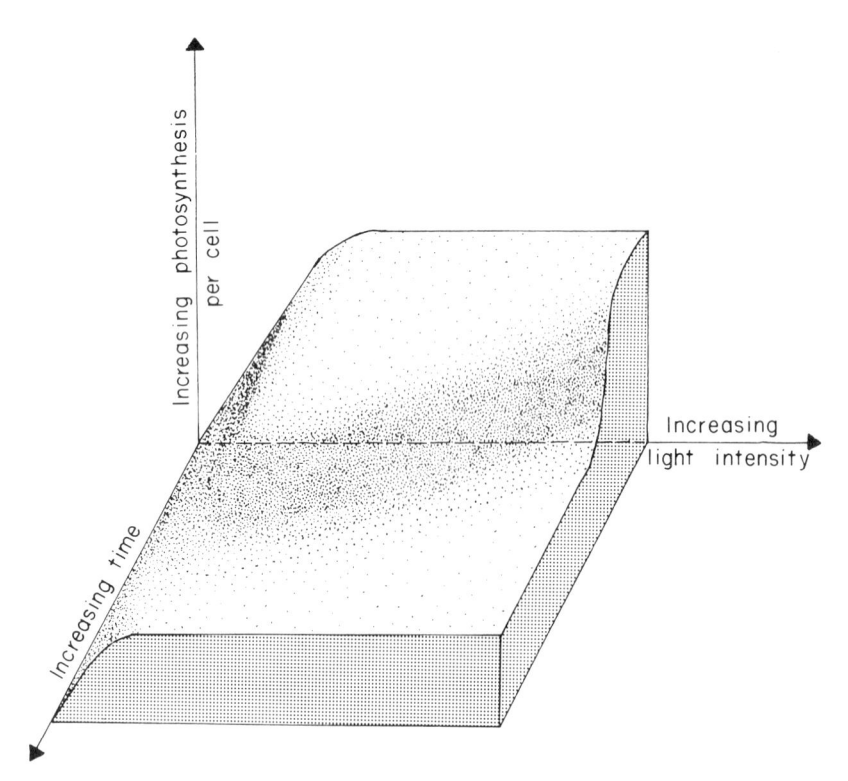

Fig. 3.6 Time-dependent changes in P versus I relationships. Cells at time 0 have been incubated in the dark for several hours. (Data from Marra 1978.)

plankton ecology because P_{max} per unit chlorophyll is one of the parameters most widely used to describe and construct mathematical models of phytoplankton production. Often called the assimilation number (mg C fixed (mg chl $a)^{-1}$ h^{-1}), values for this parameter range from 0.1 to over 20 although most are between 2 and 10. In general, low assimilation numbers are characteristic of shade-adapted or nutrient-limited cells whereas higher values are characteristic of sun-adapted or nutrient-sufficient cells. Thus, there has been a tendency to use the assimilation number as an index of the physiological state of natural phytoplankton. Care must be exercised, however. Assimilation numbers vary widely among different species of phytoplankton grown in culture at maximum growth rates and the response of the assimilation number to nutrient limitation appears to be species specific (Glover 1980). In studies of natural phytoplankton assemblages, temperature appears to be the most significant environmental parameter affecting assimilation number; at least in coastal waters, nutrient concentrations are not significantly correlated with assimilation numbers (Harrison & Platt 1980). Thus the best use of assimilation numbers as an index of physiological state appears to be for local, short-term comparisons.

Light and metabolism

Light intensity is known to affect the pattern of macromolecular synthesis from photosynthetically fixed CO_2. Whereas total CO_2 fixation declines at lower light intensities (see Fig. 3.4), the percentage of the fixed carbon which is incorporated into protein increases, while the percentage incorporated into carbohydrate decreases. At higher light intensities, carbon fixation exceeds the rate of protein synthesis (which is limited by nitrogen assimilation) and the excess carbon is stored as carbohydrate (Konopka & Schnur 1980). When cells are growing in light–dark cycles, the synthesis of protein and other macromolecules continues into the dark period at the expense of carbon and energy stored in carbohydrate. Limited evidence suggests that sun–shade adaptation does not alter the pattern of macromolecular synthesis at a given light intensity.

Although most phytoplankton species grow faster in continuous light than in light–dark cycles, a few species are sensitive to continuous light, even at low light intensities (Brand & Guillard 1981). They either grow more slowly in continuous light or they do not reproduce at all. This sensitivity is more common among oceanic species than coastal species and its biochemical basis is not known. If as suggested by some workers (see pp. 57–8), there is a selective advantage in diel phasing of cellular events, perhaps temporal separation of metabolic processes is so ingrained in sensitive species that certain processes can take place only in the dark.

Complementary chromatic adaptation

Since blue-green wavelengths dominate low-light environments in natural aquatic systems, another selective adaptation would be a shift in pigment ratios to maximize the absorption of blue-green light, i.e. increased accessory pigment/chlorophyll ratios at depth. This phenomenon, called complementary chromatic adaptation, was not observed in the 17 species of marine phytoplankton studied by Vesk and Jeffrey (1977) except for a slight increase in phycoerythrin (peak absorbance 545 nm) in the cryptomonad. Complementary chromatic adaptation is well known, however, in the blue-green algae (Bogorad 1975). Cells grown in green light (540 nm) have higher relative proportions of the red phycoerythrin (peak absorbance 540–565 nm) while cells grown in red light (640 nm) have higher relative proportions of the blue phycocyanin (peak absorbance 610–640 nm). The former response would be of obvious significance to planktonic blue-green algae which are often found at depth in the water column. The extent of the response varies significantly with the species.

NUTRIENTS

In addition to carbon, hydrogen, and oxygen, algae require some 13–15 additional elements to grow and reproduce. Most of these nutrients are usually present in sufficient amounts, relative to the alga's needs, so as not to be potential limiting factors for growth. The concentrations of nitrogen and phosphorus, however, are often low enough to limit phytoplankton growth in surface waters. Whereas phosphorus limitation is more common in lakes, nitrogen limitation is more common in the sea. Silicon concentrations can be low enough to limit diatom growth in both freshwater and marine systems.

Uptake kinetics

Nutrient uptake is thought to occur by way of specific, energy-mediated, membrane-bound, permease systems which function to provide elevated, intracellular concentrations of these ions as substrates for the various assimilatory, enzymatic processes. The uptake rate is related to the extracellular nutrient concentration by a hyperbolic function (Fig. 3.7) which is empirically described by the Michaelis–Menten expression for enzyme kinetics:

$$V = V_m \left(\frac{S}{K_s + S} \right) \quad \text{(Eqn 3.1)}$$

where V = the nutrient uptake rate, V_m = the maximum nutrient uptake rate, S = the concentration of nutrient, and K_s = the half-saturation constant or substrate concentration at which $V = V_m/2$.

By analogy with enzyme kinetics, the constant K_s is thought to reflect the affinity of the permease system for the substrate; a lower K_s suggests a higher substrate affinity and therefore an ability to take up nutrients at lower substrate concentrations. The molecular basis for the observed Michaelis–Menten kinetics of NO_3^- uptake probably resides in a membrane-

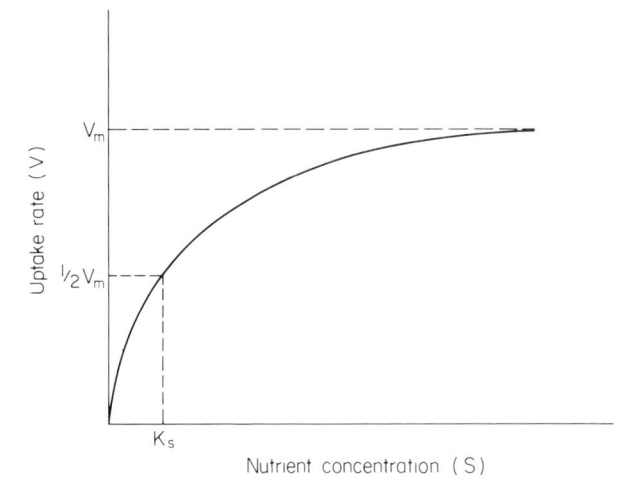

Fig. 3.7 Relationship between the substrate concentration (S) of a nutrient and the uptake rate. V_m = the maximum rate of uptake; K_s = the half-saturation constant or substrate concentration at which the uptake rate is one-half of the maximum rate.

bound ATPase activity which is stimulated by NO_3^- (Falkowski 1975). The activity, which presumably functions to transport NO_3^- across the plasma membrane with the concomitant hydrolysis of ATP, was found in marine species of five different algal classes. The K_m values for NO_3^- activation of the ATPase were in close agreement with the K_s values for uptake by whole cells.

Half-saturation constants for nitrate, ammonia, phosphate and silicate are in the $\mu mol\ l^{-1}$ range (0.1–5 $\mu mol\ l^{-1}$) which is similar to the concentrations of these nutrients frequently encountered in surface waters. Among marine phytoplankton, which are more thoroughly studied than freshwater forms, there is a definite trend in K_s toward lower values in smaller species and in species from oligotrophic waters. Intraspecific differences also have been recorded for several species: the reported K_s values for estuarine strains are up to five fold higher than those of oceanic isolates. These differences presumably reflect genetic (since they are stable in culture) adaptations to the relative ambient nutrient levels in their respective environments. It should be noted that the K_s may vary with environmental parameters such as temperature and the concentration of mono- and divalent cations.

Nutrient uptake is often, but not always, stimulated by light, sometimes as much as 10–15 fold over dark uptake rates. This finding is not unexpected for an energy-requiring process. When light stimulation does occur, the curve of uptake rate versus light intensity is not unlike the P versus I curve (see Fig. 3.4), except that the rate usually does not fall to zero in the dark. There is some evidence in the literature that light stimulation is most pronounced in nutrient-sufficient cells. Nutrient-deficient cells appear to mobilize endogenous energy reserves in order to maintain high rates into the dark period. Nutrient uptake rates are also affected by sinking rates and cell motility (see p. 59).

Nutrient-limited growth

It is now known that nutrient supply can affect the growth rate of an alga in addition to affecting its final yield. The growth rate of an alga will decline if the concentration and/or supply rate of a given nutrient drops below that supporting an uptake rate sufficient to maintain the existing growth rate. Such *nutrient-limited* growth is best studied in chemostat cultures for the reasons given in Appendix 2. *Nutrient-saturated* growth is best studied during the exponential phase of batch cultures. *Nutrient starved* refers to cells which have no further exogenous source of the nutrient in question and have stopped, or will soon stop, growing altogether. Nutrient starvation is readily achieved in the stationary phase of a batch culture or by stopping the inflow of fresh medium to a chemostat culture, thereby turning it into a batch culture rapidly approaching the stationary phase. Since it is unlikely that a population of non-growing cells will maintain itself in the euphotic zone against grazing pressure and other losses, it would seem that data on nutrient-limited cells would be more directly applicable to most natural phytoplankton populations than would data on nutrient-starved cells.

The relationship between phytoplankton growth rates and limiting nutrients usually can be described by one of two hyperbolic functions. In some studies it has been found that the same relationship used to describe uptake kinetics (see Eqn 3.1 and Fig. 3.7) also relates growth rate to the nutrient concentration in the medium. This is the Monod equation or external nutrient control model:

$$\mu = \mu_m \left(\frac{S}{K_s + S} \right) \quad \text{(Eqn 3.2)}$$

in which μ is the specific growth rate, μ_m is the maximum specific growth rate, S is the substrate concentration and K_s is the half-saturation constant.

In other studies it has been observed that growth rate is related to the cellular concentration of the limiting nutrient rather than to its external concentration. This is the Droop equation or internal nutrient control model:

$$\mu = \mu'_m \left(1 - \frac{Q_0}{Q} \right) \quad \text{(Eqn 3.3)}$$

in which Q is the cell quota of the limiting nutrient (i.e. the total amount of the limiting nutrient, in all of its forms, in the cell), Q_0 is the minimum cell quota (i.e. the subsistence quota or lowest level of Q at which the cell can grow) and μ'_m is the specific growth rate at infinite Q. These relationships are shown in diagrammatic form in Fig. 3.8.

Under steady-state conditions both equations describe nutrient-limited growth equally well because nutrient uptake (a function of external nutrient

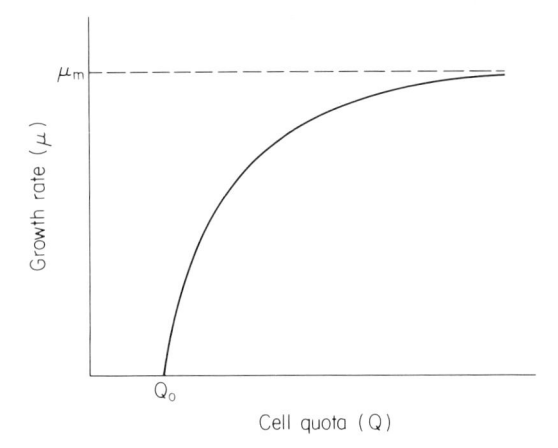

Fig. 3.8 Relationship between the specific growth rate (μ) and the cellular nutrient concentration, cell quota (Q). Q_0 = the minimum cell quota; μ_m = the maximum growth rate. (After Rhee 1980.)

concentration) is in equilibrium with the cell quota of the nutrient. Our analytical methods are not of sufficient sensitivity, however, to measure the very low, external nutrient concentrations, often associated with low growth rates in the Monod model. The concentration of the limiting nutrient in the medium of chemostat cultures, for example, is often below the limit of detection except when growth rates approach μ_m. The Droop model offers the advantage that the cell quota is readily measured in the particulate matter following concentration by filtration or centrifugation. The Droop model is constrained, however, by the fact that Q is not infinite but has an upper limit, Q_m. Thus μ_m, the true maximum growth rate, is smaller than μ'_m by the factor $(1 - Q_0/Q_m)$. For some nutrients which constitute a relatively small fraction of the total cell biomass, such as vitamin B_{12}, iron or phosphorus, there is a large difference between Q_0 and Q_m, the fraction Q_0/Q_m is small (e.g. <0.1) and μ'_m approaches μ_m. For other nutrients, such as nitrogen, which make up a large fraction of the cell biomass, the fraction Q_0/Q_m remains large and $\mu_m \ll \mu'_m$. The Droop model may be used provided that μ_m and Q_m are measured experimentally during nutrient-saturated growth. The reader should refer to Goldman and Peavey (1979) and Rhee (1980) for further discussion.

Rhee (1978) has investigated the transition state between nitrogen-limited growth and phosphorus-limited growth in *Scenedesmus*. This study addresses the question of duel nutrient limitation and also illustrates principles of chemostat operation and cell quotas. The experiment was conducted in a chemostat culture held at a constant dilution rate (and therefore a constant growth rate) in which the N : P atomic ratio of the inflowing medium was increased in stages by increasing the N concentration while the P concentration remained constant. The cell density data (Fig. 3.9a) show that the cells are N-limited up to an N : P ratio of 30 and probably P-limited above this ratio. The abrupt transition at the point where N : P = 30 suggests that the cells shift from N limitation to P limitation without being limited simultaneously by both elements during growth near the critical ratio of 30. This conclusion is supported by data on the N and P cell quotas,

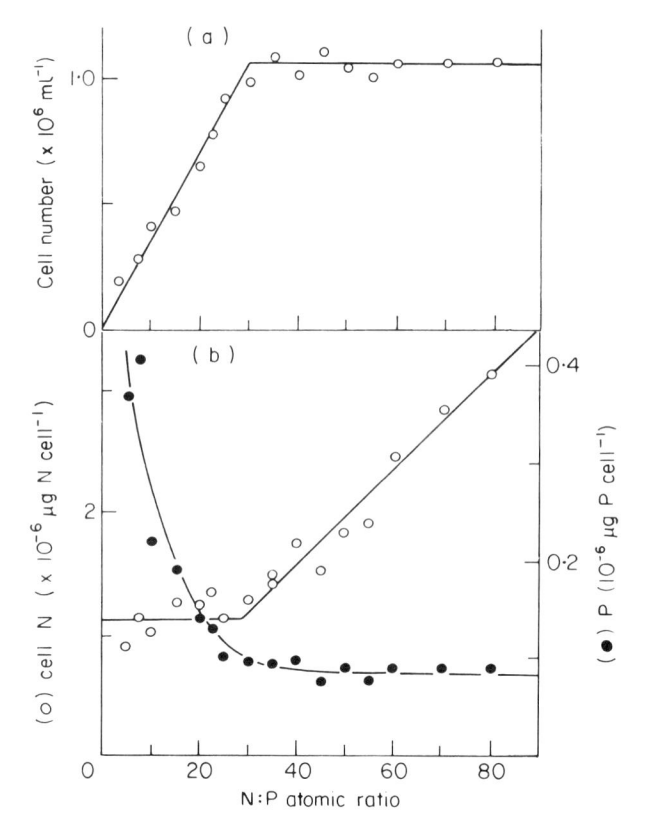

Fig. 3.9 Response of *Scenedesmus* sp. to different N : P ratios in chemostat cultures held at a single dilution rate. N : P ratios were obtained by holding the P concentration constant and varying the N concentration. There was no detectable residual N or P in the medium at any N : P ratio examined. (a) Cell density; (b) cell quota of N (open circle) and P (solid circle). (After Rhee 1978.)

Q_N and Q_P respectively (Fig. 3.9b). Below N : P = 30, Q_N is low and characteristic of cells at that N-limited growth rate. Above N : P = 30, Q_N steadily increases with increasing N in the inflow medium. Up to N : P = 30, Q_P steadily decreases because more cells are competing for the same P supply. Above N : P = 30 the cells are P-limited and Q_P is low and characteristic of P-limited cells.

Physiological adaptations to nutrient-limited growth

The most general response to increased nutrient limitation is a decrease in the cell quota of that nutrient. The cell gets by with less of the nutrient in order to continue growing. The cell quotas of non-limiting nutrients also may decline, but not nearly as much as the limiting nutrient. Carbon per cell exhibits various responses to N, P and Si limitation in different algae. Figure 3.10 illustrates how cell quotas change as a function of a N-limited growth rate in the green flagellate, *Dunaliella tertiolecta*. The sharp rise in

36

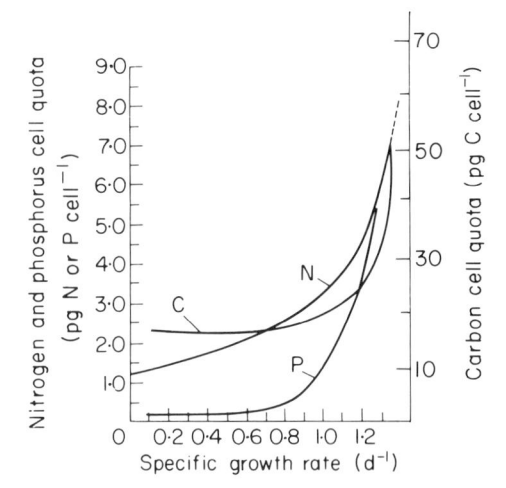

Fig. 3.10 Cell quotas of N, P and C in *Dunaliella tertiolecta* as a function of a nitrogen-limited growth rate. Maximum growth rate = 1.34 d^{-1}. (After Goldman & Peavey 1979.)

cell quotas near μ_m emphasizes the point that trends observed at a few nutrient-limited growth rates cannot uncritically be extrapolated to nutrient-starved or nutrient-saturated cells.

Other aspects of cell composition and physiology are also influenced by increasing nutrient limitation. Although many variations are observed, the general trends are a decrease in chlorophyll and protein content per cell and an increase in carbohydrate (and sometimes lipid). With respect to cell metabolism, rates of photosynthesis and respiration decline with increasing nutrient deficiency while the ability to take up the limiting nutrient is frequently enhanced. The data in Tables 3.1 and 3.2 compare a number of cellular parameters in a small, marine, centric diatom at several N- and P-limited growth rates.

In N-limited chemostat culture (Table 3.1), at the highest dilution rate (0.99 d^{-1}), the cells are growing close to their maximum rate of growth in batch culture (μ_m = 1.05 d^{-1} or generation time = 15.8 h). The generation time increases more than five fold at the lowest dilution rate where nitrogen limitation becomes severe. Cell concentration in the culture increases with increasing N limitation (as explained on p. 161) while cell volume as well as the cell quotas of carbon, nitrogen, phosphorus and chlorophyll *a* decline. Nitrogen, phosphorus and chlorophyll *a* decline faster than carbon as shown by the increasing C : N, C : P and C : chl *a* ratios. Cell metabolism also slows down as shown by decreases in the light-saturated rate of photosynthesis, V_m for PO_4^{-3} uptake, and in the activity of glutamic dehydrogenase. Although the half-saturation constants for NO_3^- and NH_4^+ uptake (not shown) did not show any consistent trends, the maximum uptake rates for these ions increase as nitrogen limitation becomes more severe. The increase in the specific rate of nitrogen uptake (N taken up per unit cell N per unit time) is more striking because of the decrease in the cell quota of nitrogen. The specific rate is significant because it reflects the cell's ability to acquire that amount of N needed in order to reproduce.

Table 3.1 Cellular characteristics of *Thalassiosira pseudonana* under nitrogen-limited growth. The chemostat was maintained in a **12** h light–12 h dark cycle with equimolar levels of NO_3^- and NH_4^+ in the dilution medium. (Data from Eppley & Renger 1974 and Perry 1976.)

	Dilution rate (d^{-1})		
	0.99	0.42	0.20
Generation time (days)	0.70	1.65	3.47
Cell concentration (10^3 ml^{-1})	177	558	605
Cell volume (μm^3)	120	80	102
Carbon (pg $cell^{-1}$)	14.0	9.0	8.0
Nitrogen (pg $cell^{-1}$)	2.74	1.05	0.66
Phosphorus (pg $cell^{-1}$)	0.56	0.28	0.23
Chlorophyll (pg $cell^{-1}$)	0.49	0.20	0.09
C : N (g : g)	5.1	8.6	12.1
C : P (g : g)	24.6	32.7	35
C : chl a (g : g)	28.9	44.3	90.9
P_{max} (pg C $cell^{-1}$ h^{-1})	2.09	1.01	0.44
Assimilation number (mg C × mg chl a^{-1} h^{-1})	5.0	4.6	4.4
$V_m(NO_3^- - N)$ (pg $cell^{-1}$ h^{-1})	0.24	0.25	0.38
$V_m(NO_3^-)$ h^{-1} (specific rate)	0.088	0.24	0.58
$V_m(NH_4^+ - N)$ (pg $cell^{-1}$ h^{-1})	0.27	0.31	0.39
$V_m(NH_4^+)$ h^{-1} (specific rate)	0.099	0.30	0.59
$V_m(PO_4^{-3} - P)$ (pg $cell^{-1}$ h^{-1})	2.40	0.33	0.71
$V_m(PO_4^{-3})$ h^{-1} (specific rate)	4.29	1.18	3.09
Glutamic dehydrogenase*	31	7.6	3.1

* = 10^{-15} g at N $cell^{-1}$ h^{-1}.

The isolate of *Thalassiosira pseudonana* studied in P-limited chemostat culture (Table 3.2) was smaller. Note that with increasing PO_4^{-3} deficiency the cell quota of nitrogen remains constant while that of phosphorus drops. Chlorophyll per cell declines as it did with N deficiency, but the C per cell increases in contrast to a decrease in C per cell experienced during N-

Table 3.2 Cellular characteristics of *Thalassiosira pseudonana* under phosphorus-limited growth. The chemostat was maintained in a 15 h light–9 h dark cycle. (Data from Perry 1976.)

	Dilution rate (d^{-1})	
	0.99	0.42
Generation time (days)	0.70	1.65
Cell concentration (10^3 ml^{-1})	400	670
Carbon (pg $cell^{-1}$)	3.0	5.4
Nitrogen (pg $cell^{-1}$)	0.60	0.59
Phosphorus (pg $cell^{-1}$)	0.034	0.021
Chlorophyll (pg $cell^{-1}$)	0.105	0.081
C : N (g : g)	5.0	9.2
C : P (g : g)	88	257
C : chl a (g : g)	28.6	66.7
$V_m(PO_4^{-3} - P)$ (pg $cell^{-1}$ h^{-1})	1.38	1.41
$V_m(PO_4^{-3})$ h^{-1} (specific rate)	40.6	67.1

limitation. The ratios C : N, C : P and C : chl *a* all increase as they did with N-deficiency except that C : P increases more dramatically under P limitation than C : N did under N limitation. The maximum specific PO_4^{-3} uptake rate is also much higher in P-limited conditions than in N-limited cells.

It is perhaps unexpected that the enhancement of nutrient uptake observed during nutrient-limited growth is expressed as an increase in V_m (the ability to take up nutrients at high nutrient concentrations) rather than a decrease in K_s (a greater affinity for the limiting nutrient). In other words, the cell apparently allocates resources to synthesizing, additional, uptake sites to increase maximum uptake rates, but does not modify the uptake mechanism *per se* to more efficiently utilize the ambient supply (as perceived by currently available techniques). In fact, under nutrient-limited conditions, V_m can be many times higher than the uptake rate required to support the low rates of cell replication observed under these conditions. McCarthy and Goldman (1979) suggest that this paradox is the result of a problem of scale. Phytoplankton cells are responding to volumes in the order of a μl, whereas nutrient analyses usually are conducted on volumes of 50 ml (obtained from a larger, mixed-water sample). Thus any micropatches of nutrient-rich water which may exist as a result of zooplankton excretion or bacterial remineralization are undetectable. It is not unlikely that the high V_m is an adaptation to immediately and rapidly assimilate limiting nutrients from these nutrient-rich microhabitats (perhaps only nanolitres in size). Thus, uptake at V_m in a nutrient-rich patch, for a small fraction of the cell's doubling time, would support growth in an otherwise nutrient-depleted environment.

All phytoplankton species are not equally capable of adapting to low nutrient conditions. The short-term, physiological adaptations discussed above are superimposed on more rigidly fixed, evolutionary adaptations to various nutrient conditions. As discussed on p. 33, for example, open-ocean strains of some diatom species have lower K_s values than coastal strains. Although much more research is needed, our present understanding of the comparative nutrient relationships of species from nutrient-poor and nutrient-rich environments is summarized in Fig. 3.11. Species evolutionarily adapted to oligotrophic environments tend to do better at low nutrient concentrations but have lower nutrient-saturated rates than species from eutrophic environments.

Nutrient status of natural phytoplankton

Are natural phytoplankton assemblages nutrient-limited? If so, which nutrient is limiting and how severe is the limitation? These straightforward and important questions are not easy to answer. The standing stock of a nutrient in solution is generally an inadequate indicator of nutrient status. Whereas it is usually safe to assume that nutrient concentrations above the $\mu mol\ l^{-1}$ range (e.g. 1 $\mu mol\ l^{-1}\ PO_4^{-3}$ and 10 $\mu mol\ l^{-1}\ NH_4^+$) are not limiting, lower, even very low, concentrations may or may not be limiting depending on the

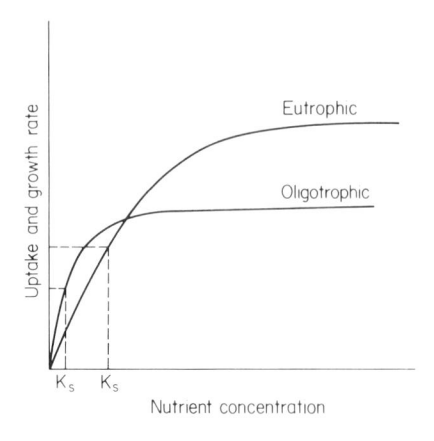

Fig. 3.11 Comparison of the substrate-dependent uptake and growth rates in phytoplankton evolutionarily adapted to oligotrophic and eutrophic conditions.

presence of other limiting factors and on the rate at which the nutrient is recycled in that system. As mentioned above, relatively high growth rates may be obtained in chemostats when the nutrient is below the level of detection in the outflow medium. Even non-limiting nutrients may be at undetectable levels (Rhee 1978 and Fig. 3.9). Furthermore, different species will become limited at various nutrient concentrations.

Cell composition has the potential to be a useful indicator of nutrient limitation because of the significant responses to the nutrient-limited growth rates discussed above. Cells growing near nutrient-saturated growth rates typically contain C, N and P in the atomic ratio of 106 : 16 : 1, the Redfield ratio (Goldman et al. 1979). (This is the reason many culture media contain an N : P ratio of approximately 15.) Therefore, a high N : P ratio (e.g. 30) in a phytoplankton sample would suggest phorphorus limitation while a low ratio (e.g. 5) would suggest nitrogen limitation. Healey and Hendrel (1980) have identified a number of cell parameter ratios in cultured algae which are sensitive to nutrient deficiency. Due to large variations between species these ratios only serve as approximate indices of the nutrient status of natural phytoplankton, but they did show clearly the expected increase in P deficiency with an increasing N : P loading ratio in a series of Canadian experimental lakes (Fig. 3.12). The influence of contamination by non-living particulate matter on these estimates apparently is not as great as might be imagined because detritus tends to have an elemental composition similar to that of living cells. Those ratios involving metabolic (phosphatase activity, P debt) or labile (ATP) components are less subject to interference by detritus.

A nutrient-enrichment bio-assay is the most commonly employed method for identifying limiting nutrients in natural phytoplankton communities. Nutrients are added to natural phytoplankton samples and their effect is measured as an increase in cell number, chlorophyll or productivity. For example, in a nitrogen-limited situation, samples enriched with NH_4^+ or NO_3^- or with a mixture of nutrients including N, would exhibit enhanced growth compared with samples enriched with non-nitrogenous single nutri-

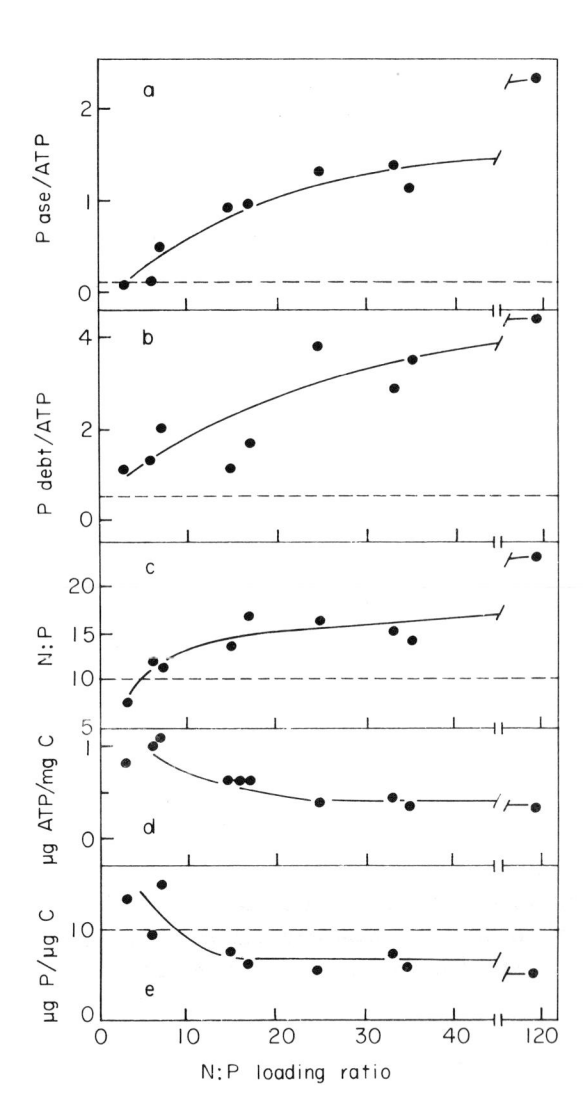

Fig. 3.12 Physiological indicators of phosphorus deficiency in phytoplankton from a series of experimental lakes with increasing N : P loading ratios (by weight). Values above the dashed line in (a), (b) and (c) indicate P deficiency; all values shown in (d) indicate P deficiency; and values below the dashed line in (e) indicate severe P deficiency. P ase = alkaline phosphatase (see p. 42); P debt = the amount of P taken up during 24–36 hours of darkness. (After Healey & Hendzel 1980.)

ents or with a nutrient mixture lacking N. Cell number and chlorophyll measurements require incubation periods of several days before significant differences can be expected. Productivity measurements with ^{14}C-HCO$_3^-$ can be misleading if carried out shortly after enrichment because nutrient uptake and assimilation will be competing with CO_2 fixation for metabolic energy (ATP). Since a nutrient-limited cell has a greater immediate need for a limiting nutrient than for additional carbon, photosynthesis is often reduced rather than enhanced for up to a day following enrichment (Gerhart & Likens 1975). Many changes in water chemistry and species composition which occur during this period could complicate the interpretation of results. These problems are minimized by using *in situ* 'big bag' enclosures (see p. 161), but this method is not suitable for routine work. Another approach to the nutrient-enrichment bio-assay is to inoculate filtered water with a culture of a test alga. The response of a single test species, however, may not be representative of the natural assemblage.

Our lack of understanding of the nutrient status of natural phytoplankton is illustrated by the current debate over the growth rates of phytoplankton in the nitrogen-limited central gyres of the oceans. One group (Goldman *et al.* 1979) argues that the algae are growing close to the maximum rate (by utilizing micropatches of nutrient-rich water) since the C : N : P ratio of the particulate matter is within the range expected of cells growing near μ_m. Another group (Sharp *et al.* 1980) argues that the algae are growing slowly because the C and N turnover times are low.

Nitrogen

Most algae can utilize NO_3^-, NO_2^- or NH_4^+ as nitrogen sources, although some flagellates, especially euglenoids, cannot grow on NO_3^- or NO_2^-. Nitrite is not as abundant in natural waters as the other forms of fixed nitrogen and, in fact, may be toxic if present at the levels normally employed in culture media. The ability to use various forms of organic nitrogen, particularly urea, is widespread among the algae. Ammonium is usually preferred over NO_3^- to the extent that NH_4^+ concentrations above 0.5–1.0 μmol l^{-1} will inhibit uptake of NO_3^-. Algae presumably conserve energy by utilizing NH_4^+ because NO_3^- must be reduced by the cell to NH_4^+ before incorporation into amino acids. The advantage, if it exists, is not evident as increased growth rates in continuous light, although increased growth rates on NH_4^+ have been reported in light-limited situations. The specific activity of nitrate reductase, the enzyme which reduces NO_3^- to NO_2^-, is very low or undetectable in cells growing on NH_4^+ or in nitrogen-starved cells, but increases rapidly if cells are given NO_3^- as a sole nitrogen source. Hence, the specific activity of this enzyme has been used as an indicator of nitrate utilization by natural phytoplankton communities.

Phosphorus

Orthophosphate, PO_4^{-3}, is the only important inorganic phosphorus source for algae although most can obtain the element from various organic phosphates. Glycerophosphate, for example, is often used in culture media. Membrane-bound, alkaline and acid phosphatases hydrolyse these compounds to liberate PO_4^{-3} which is then taken into the cell. Under conditions of PO_4^{-3} deficiency, alkaline phosphatase activity usually increases dramatically (10–100 fold), an apparent adaptive strategy to P limitation. Most algae store excess PO_4^{-3} as polyphosphate in cytoplasmic granules of 30–500 nm diameter. Polyphosphate acts as a P reservoir, decreasing in cells as a function of P-limited growth rate in much the same way that the total cell quota of P declines. The term luxury consumption is often applied to this ability of cells to accumulate a nutrient beyond their immediate needs.

Silicon

The presence of the siliceous frustule gives diatoms an unusual and absolute nutritional requirement for silicon in the form of orthosilicic acid, $Si(OH)_4$. In addition to the Si requirement for cell-wall formation, diatoms require small amounts of Si for net DNA synthesis, an unusual requirement which resides in the translation step in the synthesis of DNA polymerase. The adaptive significance of the latter Si requirement is not clear. Although other algae, particularly chrysophytes, produce siliceous scales and/or cysts, diatoms were thought to be unique among the algae in requiring Si. Recently, however, the colonial chrysophyte, *Synura petersenii*, which produces siliceous scales, was found to require at least 1 μmol l^{-1} Si for optimum growth and to deplete silicic acid from the medium, although an absolute requirement for growth was not demonstrated. The green flagellate *Platymonas* sp. also accumulates silicon but has no known silicon requirement (Fuhrman *et al.* 1978). The ecological consequences of this competition for silicic acid on diatom growth and on patterns of seasonal succession remain to be determined.

Studies with synchronized diatom cultures have shown that $Si(OH)_4$ uptake is largely confined to that portion of the cell cycle in which the two daughter cells are each synthesizing a new valve. In other words, the cell does not accumulate and store Si during the early part of the cell cycle to use later during wall formation. In at least one diatom uptake rate increases ten fold during the cell cycle, with a peak during wall formation, due to increases in K_s and V_m (Sullivan 1977).

Table 3.3 contains data on the cell composition of the common, marine, centric diatom, *Skeletonema costatum*, at different levels of silicon supply. These data should be compared with those in Tables 3.1 and 3.2. Note that Si per cell declines with Si limitation as is typical of the limiting nutrient, while the cellular levels of other components increase. Various workers have observed thinner cell walls and shorter spines in diatoms under Si-limited

Table 3.3 Comparison of cell composition in *Skeletonema costatum* under different levels of silicon supply. Data on Si-saturated and Si-starved cells were obtained from batch culture; data on Si-limited cells from chemostat culture. (From Harrison *et al.* 1977.)

	Growth conditions		
	Si-saturated	Si-limited	Si-starved
Cell volume (μm^3)	181	133	150
Carbon (pg cell^{-1})	8.5	13.0	28.5
Nitrogen (pg cell^{-1})	1.94	2.23	2.52
Phosphorus (pg cell^{-1})	0.434	0.588	0.709
Silicon (pg cell^{-1})	2.11	1.09	1.81
Chlorophyll *a* (pg cell^{-1})	0.14	0.20	0.16
C : N (g : g)	4.37	5.83	11.3
C : P (g : g)	19.7	22.1	40.2
C : Si (g : g)	4.03	11.9	15.7
C : chl *a* (g : g)	61	65	178

growth. The unexpected increase in Si per cell in Si-starved cells relative to Si-limited cells (Table 3.3) is probably related to the fact that they were produced in batch culture. When a Si-saturated culture rather abruptly enters the Si-starved stationary phase, there is insufficient time for thinner frustules to be produced. The ratios $C:N$, $C:P$, $C:Si$ and $C:chl\ a$ all increase in Si-limited cells as they do in N- and P-limited cells. Interestingly, Si-starved cells accumulate lipid as a reserve product whereas N- and P- starved diatoms accumulate carbohydrate (data not shown). As a final note, the observations that silicic acid uptake is largely confined to one part of the cell cycle and that Si starvation serves to synchronize the cell cycle, suggest that any measurement of cell parameters in Si-limited or Si-starved diatoms should include an analysis of the cell cycle. The question is, Are the effects observed a direct result of Si depletion or are they an indirect result caused by an accumulation of cells in a certain phase of the division cycle?

Trace elements

Although not as commonly implicated to be limiting factors as N, P, and Si, other inorganic elements sometimes may be present at limiting concentrations, either as a deficiency or at toxic concentrations. Elements which are required in very small amounts (trace elements) are usually present at sufficient levels in eutrophic waters, but may be deficient in oligotrophic waters. Enrichment experiments have demonstrated that productivity is enhanced by the addition of Fe, Mn, Mo, Co, and Zn in various aquatic systems (e.g. Axler *et al.* 1980). Trace-element nutrition is a complicated subject because the absolute concentration is often less important than the 'availability' of the element, which can be affected by a number of environmental variables. Organic ligands (chelating agents) are especially important in that they form dissociable complexes with many metal ions, helping to keep them in solution and to maintain concentrations of the free ion at optimum levels. In a sense, they act as metal ion buffers. Ethlenediaminetetraacetic acid (EDTA) is a synthetic chelating agent used in culture media for this purpose, but many naturally occurring organic molecules, especially 'humic' substances, are thought to function in a similar capacity. The single addition of EDTA to natural water can enhance productivity either by increasing the availability of trace metals which are already present or by decreasing toxic levels of free ions. The addition of EDTA may also decrease productivity by over-chelating the trace metal, thus reducing the level of the free ion to limiting levels.

Vitamins

If an alga requires an exogenous source of a vitamin, it will be one or more of the following three: vitamin B_{12}, thiamine, or biotin. In some groups of algae more than half of the species studied may require vitamins (e.g. Pyrrhophyceae, Euglenophyceae, Chrysophyceae) while in other groups (e.g. Cyanophyceae, Chlorophyceae) less than half require vitamins. Vitamin concentrations in natural waters range from tenths to hundreds of ng l^{-1} (Provasoli & Carlucci 1974). The only practical way to measure such low concentrations is to use vitamin-requiring bacteria or algae as test organisms in bio-assays. Algae also secrete into the medium vitamins which they are able to synthesize endogenously. In fact, it is possible to grow a vitamin-requiring species in an initially vitamin-free culture if a vitamin-secreting species is included. Since vitamin requirements are not universal among algae, vitamin supplies are more likely to affect species composition of the phytoplankton than its overall productivity.

A number of different algae produce a glycoprotein that binds free vitamin B_{12} and renders it unavailable to vitamin B_{12}-requiring algae (Pintner & Altmeyer 1979). In dense cultures, the binder is released into the medium. It is not yet clear whether the binder functions primarily as a cellular B_{12}-trapping molecule which is incidentally released to the medium or whether it functions in competitive interactions by inhibiting the growth of vitamin B_{12}-requiring algae. It is interesting that the substance is produced by both B_{12}-producing and B_{12}-requiring algae.

TEMPERATURE

The third major environmental factor to affect phytoplankton growth is temperature. Since the temperature of large bodies of water is relatively constant on a diel basis and is usually below 30°C, phytoplankton ecologists are more concerned with the seasonal and latitudinal impact of temperature and with long-term adaptation to temperature rather than with the impact of daily temperature fluctuations. One of the most important effects of temperature on phytoplankton is indirect through its impact on water column stability (p. 23).

Phytoplankton organisms exhibit the usual relationship between temperature and biological activity by increasing growth rate with increasing temperature up to some optimum temperature after which growth rate declines, often abruptly, to zero (see Fig. 3.14a). The temperature optima for growth of many marine and freshwater phytoplankters lie in the range 18–25°C, although cold-water forms generally have lower optima. Some Antarctic diatoms have temperature optima of 4–6°C and will not grow above 7–12°C, whereas an isolate of a tropical diatom will not grow below 17°C. Algae are able to adapt, within limits, to suboptimum temperatures, but the mechanisms are not well understood. In general, algae with higher optimum temperatures for growth have higher growth rates at their optimum temperatures than algae with lower temperature optima for growth. It is interesting that the optimum temperature in culture is usually higher than the maximum temperature at which the organism is found growing in nature. The observation does not necessarily mean that the algae are temperature limited beginning to uncover the much more complex, and apparently species specific, relationships between temperature and cell composition, particularly in nutrient-limited and light-limited situations. The Droop model of nutrient-limited growth, in particular, will be affected by these discoveries because growth rate is predicted on the basis of cell quota.

SALINITY

Dissolved inorganic salts in marine and fresh waters may potentially affect phytoplankton growth as a function of their elemental composition or as a function of their osmotic activity. The major cations and anions involved (Na^+, K^+, Mg^{++}, Ca^{++}, HCO_3^-, $CO_3^=$, Cl^-, $SO_4^=$), are present in high enough concentrations that they are not likely to be limiting in the sense that N, P or Si are often limiting. In the sea where the ratios of major ions are remarkably constant, variations in elemental composition are probably not significant to phytoplankton growth. It is interesting that many marine algae, fungi and bacteria have a specific, non-osmotic requirement for millimolar levels of Na^+. This requirement is related to membrane transport phenomena, at least in marine bacteria.

In lakes the concentrations and ratios of ions vary widely. In the extremes, hard waters tend to have higher total concentrations of salts dominated by Ca^{++}, HCO_3^- and $CO_3^=$, lower monovalent/divalent cation ratios and higher pH values (above 8) whereas soft waters tend to have lower total dissolved salts with relatively higher concentrations of Na^+ and Cl^-, higher monovalent/divalent cation ratios and more acidic pH values (below 7). Distinct differences in algal floras between lakes are frequently correlated with these chemical characteristics of the water, especially the calcium level. In a detailed comparison of algal species characteristically found in soft and hard waters, Moss (1973) concluded that the level of free CO_2 dissolved in the water was the major factor influencing the observed distribution patterns. The correlation with water hardness occurs because the availability of free CO_2 decreases as the pH increases with increasing hardness. Species growing in hard waters are able to assimilate HCO_3^- as a carbon source whereas those growing in soft waters are probably restricted to using free CO_2 as a carbon source.

The osmotic activity of dissolved solids also has important consequences for the growth and distribution of phytoplankton. Many species from soft-water lakes require total salt concentrations below 100–200 parts/10^6 (0.1–0.2‰). Some species from estuarine habitats can grow well in salinities ranging from fresh water to full strength sea-water (35‰) or above, whereas other species are less tolerant, requiring brackish water with more limited salinity ranges (e.g. 4–20‰). Phytoplankton from coastal waters usually have salinity optima for growth in culture below full strength sea-water, perhaps 20–25‰. Oceanic species appear to do best at full strength sea-water, but few have been examined for salinity responses. In a special case, diatoms forming blooms in the brine solutions on the underside of Arctic sea ice will grow at salinities ranging from 5‰ to 60‰ (Grant & Horner 1976).

In general, freshwater algae have an osmotic potential equivalent to about 0.15M NaCl, significantly above that of most freshwater environments. Water tends to enter these cells by osmosis. Cells which lack cell walls or have weak cell walls, usually have contractile vacuoles to get rid of excess water, whereas cells with stronger cell walls use turgor pressure to counteract net water influx. Marine algae also tend to have higher osmotic potentials than their environment, which is equivalent to about 0.5M NaCl. Most of them appear to accumulate K^+ and exclude Na^+ relative to the concentrations of these ions in sea-water.

Phytoplankton cells adapt to the osmotic stress of changing salinities by adjusting their internal solute concentration in order to return to equilibrium with the environment (Hellebust 1976a). Since cytoplasm occupies more of the cellular volume than does the vacuole in most of these small cells, organic solutes which are compatible with metabolic processes are the principal osmoticum employed. A variety of solutes are synthesized, including polyols such as glycerol, mannitol and sorbitol as well as proline, mannose, sucrose and isofloridoside with no apparent phylogenetic pattern. Potassium and possibly other inorganic ions, which may be sequestered in the vacuole, are also involved in osmoregulation, especially in walled cells with large vacuoles.

Osmoregulation has been most intensively studied in several naked, green flagellates (e.g. certain species of *Dunaliella* and *Platymonas*) which are extremely euryhaline. In response to a sudden increase in the ionic concentration in their environment, these algae rapidly lose water and shrink, concentrating their internal solutes in the process. Over the next several hours they synthesize various polyols and slowly regain their original size. When the cells are exposed to a sudden decrease in the salinity of their medium they do not burst because the solutes rapidly leak out of the cell before water influx bursts the cell membrane. Recovery, including the return of normal membrane permeability, follows in several hours provided the shock has not been too great. In response to less severe dilution, the organic solutes are metabolized until the proper osmotic balance is achieved. The success of this osmoregulatory system is demonstrated in the ability of *Platymonas suecica* to grow at salinities ranging from 10% to 300% of normal sea-water (Hellebust 1976b).

HETEROTROPHIC GROWTH

A number of planktonic algae are capable of heterotrophic growth in laboratory culture on glucose, acetate, lactate or various other organic substrates. Thus, heterotrophy is frequently cited as a possible explanation for the occurrence of healthy algal cells far below the euphotic zone and as a means by which phytoplankton near the compensation depth could supplement their low rates of autotrophic carbon assimilation. Direct experimental support for these hypotheses, however, has been difficult to obtain. Although these algae grow well in culture at millimolar levels of organic substrates, their ability to assimilate significant amounts at the micromolar levels found in natural waters has been questioned. Vincent and Goldman (1980) have listed the criteria which must be met to conclusively demonstrate algal heterotrophy in nature: '1. Analytical description of the ambient concentrations and rate of supply of utilizable substrates. 2. *In situ* evidence that the algae have high-affinity transport systems which enable them to take up organic compounds effectively at the ambient concentrations. 3. Evidence that the algae are physiologically equipped to use these substrates for growth. 4. An assessment of the relative contribution *in situ* of dissolved organic substances versus light and CO_2 as carbon and energy sources for phytoplankton growth.'

Their studies (Vincent & Goldman 1980) in oligotrophic Lake Tahoe (California and Nevada, USA) have provided some of the best evidence to date that heterotrophy is significant to the carbon balance of phytoplankton living near the bottom of the euphotic zone. In light–dark bottle incubations, *in situ* light intensities throughout the euphotic zone stimulated the uptake of ^{14}C-acetate (0.1 μmol l^{-1}) in water samples obtained from near the bottom of the euphotic zone, but did not stimulate uptake in samples obtained at shallower depths. Photosynthetic inhibitors eliminated light-stimulated acetate uptake but had no effect on dark uptake. These and other experiments suggest that acetate uptake by phytoplankton (light-stimulated uptake) is superimposed on bacterial uptake (dark uptake) and is most active at light-limiting depths where it would be a selective advantage. Microautoradiographic analysis confirmed that natural populations of two, deep-living species of small green algae took up acetate at the ambient concentrations of acetate found in the lake. Four other species did not take up acetate. In culture both of the acetate-utilizing species grew slowly in the dark on 4 μmol l^{-1} acetate. Activities of enzymes involved in acetate metabolism were present at high levels in the water column near the bottom of the euphotic zone. Calculations based on the available data suggested that acetate uptake by these two populations was 25–247% of their photosynthetic carbon uptake. The data support the conclusion that these algae are using heterotrophically acquired carbon to supplement their autotrophic nutrition in dim light at the bottom of the euphotic zone and to provide for maintenance metabolism below the euphotic zone.

ORGANIC EXCRETION

Phytoplankton cells release soluble organic matter into the medium, mostly in the form of glycolic acid, other organic acids, simple sugars and sugar phosphates, amino acids and polysaccharides. Reports on the amount of carbon excreted vary from a few per cent to over 90% of the total carbon fixed. A number of technical problems as well as shock treatments (e.g. sudden exposure to high light intensities) can lead to high rates of excretion

(Mague *et al.* 1980). In general, lower values seem to be more typical of healthy cells when experiments are performed carefully. Over a 24 h period, the amount of carbon lost through organic excretion is likely to be somewhat similar to that lost in dark respiration. The functional significance, if any, of this phenomenon to the algae is not known. It is possible that some of the excreted molecules function as chelating agents or in allelopathic interactions (see Chapter 4).

REFERENCES

AXLER R.P., GERSBERG R.M. & GOLDMAN C.R. (1980) Stimulation of nitrate uptake and photosynthesis by molybdenum in Castle Lake, California. *Can. J. Fish. aquat. Sci.*, **37**, 707–12.
BELAY A. & FOGG G.E. (1978) Photoinhibition of photosynthesis in *Asterionella formosa* (Bacillariophyceae). *J. Phycol.*, **14**, 341–7.
BOGORAD L. (1975) Phycobiliproteins and complementary chromatic adaptation. *Ann. Rev. plant Physiol.*, **26**, 369–401.
BRAND L.E. & GUILLARD R.R.L. (1981) The effects of continuous light and light intensity on the reproduction rates of twenty-two species of marine phytoplankton. *J. exp. mar. Biol. Ecol.*, **50**, 119–32.
EPPLEY R.W. & RENGER E.H. (1974) Nitrogen assimilation of an oceanic diatom in nitrogen-limited continuous culture. *J. Phycol.*, **10**, 15–23.
FALKOWSKI P.G. (1975) Nitrate uptake in marine phytoplankton: Comparison of half-saturation constants from seven species. *Limnol. Oceanogr.*, **20**, 412–17.
FALKOWSKI P.G. & OWENS T.G. (1980) Light-shade adaptation, two strategies in marine phytoplankton. *Plant Physiol.*, **66**, 592–5.
FUHRMAN J.A., CHISHOLM S.W. & GUILLARD R.R.L. (1978) Marine alga *Platymonas* sp. accumulates silicon without apparent requirement. *Nature (Lond.)*, **272**, 244–6.
GERHART D.Z. & LIKENS G.E. (1975) Enrichment experiments for determining nutrient limitation: Four methods compared. *Limnol. Oceanogr.*, **20**, 649–53.
GLOVER H.E. (1980) Assimilation numbers in cultures of marine phytoplankton. *J. plankton Res.*, **2**, 69–79.
GOLDMAN J.C. & MANN R. (1980) Temperature-influenced variations in speciation and chemical composition of marine phytoplankton in outdoor mass cultures. *J. exp. mar. Biol. Ecol.*, **46**, 29–39.
GOLDMAN J.C., MCCARTHY J.J. & PEAVEY D.G. (1979) Growth rate influence on the chemical composition of phytoplankton in oceanic waters. *Nature (Lond.)*, **279**, 210–15.
GOLDMAN J. C. & PEAVEY D. G. (1979) Steady-state growth and chemical composition of the marine chlorophyte *Dunaliella tertiolecta* in nitrogen-limited continuous cultures. *Appl. environ. Microbiol.*, **38**, 894–901.
GRANT W.S. & HORNER R.A. (1976) Growth responses to salinity variation in four arctic ice diatoms. *J. Phycol.*, **12**, 180–5.
HARRISON P.J., CONWAY H.L., HOLMES R.W. & DAVIS C.O. (1977) Marine diatoms grown in chemostats under silicate or ammonium limitation. III. Cellular chemical composition and morphology of *Chaetoceros debilis*, *Skeletonema costatum*, and *Thalassiosira gravida*. *Mar. Biol.*, **43**, 19–31.
HARRISON W.G. & PLATT T. (1980) Variations in assimilation number of coastal marine phytoplankton: Effects of environmental co-variates. *J. plankton Res.*, **2**, 249–60.
HEALEY F.P. & HENDZEL L.L. (1980) Physiological indicators of nutrient deficiency in lake phytoplankton. *Can. J. Fish. aquat. Sci.*, **37**, 442–53.
HELLEBUST J.A. (1976a) Osmoregulation. *Ann. Rev. plant Physiol.*, **27**, 485–505.
HELLEBUST J.A. (1976b) Effect of salinity on photosynthesis and mannitol synthesis in the green flagellate *Platymonas suecica*. *Can. J. Bot.*, **54**, 1735–41.
KONOPKA A. & SCHNUR M. (1980) Effect of light intensity on macromolecular synthesis in cyanobacteria. *Microbial Ecol.*, **6**, 291–301.

MAGUE T.H., FRIBERG E., HUGHES D.J. & MORRIS I. (1980) Extracellular release of carbon by marine phytoplankton; a physiological approach. *Limnol. Oceanogr.*, **25**, 262–79.

MARRA J. (1978) Effect of short-term variations in light intensity on photosynthesis of a marine phytoplankter: A laboratory simulation study. *Mar. Biol.*, **46**, 191–202.

MCCARTHY J.J. & GOLDMAN J.C. (1979) Nitrogenous nutrition of marine phytoplankton in nutrient-depleted waters. *Science*, **203**, 670–2.

MORRIS I. (ed.) (1980) *The Physiological Ecology of Phytoplankton*. Blackwell Scientific Publications, Oxford.

MOSS B. (1973) The influence of environmental factors on the distribution of freshwater algae: An experimental study. IV. Growth of test species in natural lake waters, and conclusion. *J. Ecol.*, **61**, 193–211.

PARSONS T.R., TAKAHASHI M. & HARGRAVE B. (1977) *Biological Oceanographic Processes*, 2nd edn. Pergamon Press, Oxford.

PERRY M.J. (1976) Phosphate utilization by an oceanic diatom in phosphorus-limited chemostat culture and in the oligotrophic waters of the central North Pacific. *Limnol. Oceanogr.*, **21**, 88–107.

PINTNER I.J. & ALTMEYER V.L. (1979) Vitamin B_{12}-binder and other algal inhibitors. *J. Phycol.*, **15**, 391–8.

PLATT T. & JASSBY A.D. (1976) The relationship between photosynthesis and light for natural assemblages of coastal marine phytoplankton. *J. Phycol.*, **12**, 421–30.

PROVASOLI L. & CARLUCCI A.F. (1974) Vitamins and growth regulators. In Stewart W.D.P. (ed.), *Algal Physiology and Biochemistry*, pp. 741–87. Blackwell Scientific Publications, Oxford.

RHEE G-Y. (1978) Effects of N:P atomic ratios and nitrate limitation on algal growth, cell composition, and nitrate uptake. *Limnol. Oceanogr.*, **23**, 10–25.

RHEE G-Y. (1980) Continuous culture in phytoplankton ecology. In Droop M.R. & Jannasch H.W. (eds), *Advances in Aquatic Microbiology* **2**, pp. 151–203. Academic Press, New York.

SENGER H. & FLEISCHHACKER P. (1978) Adaptation of the photosynthetic apparatus of *Scenedesmus obliquus* to strong and weak light conditions. I. Differences in pigments, photosynthetic capacity, quantum yield and dark reactions. *Physiologica Plantarum*, **43**, 35–42.

SHARP J.H., PERRY M.J., RENGER E.H. & EPPLEY R.W. (1980) Phytoplankton rate processes in the oligotrophic waters of the central North Pacific Ocean. *J. plankton Res.*, **2**, 335–53.

SULLIVAN C.W. (1977) Diatom mineralization of silicic acid. II. Regulation of $Si(OH)_4$ transport rates during the cell cycle of *Navicula pelliculosa*. *J. Phycol.*, **13**, 86–91.

VESK M. & JEFFREY S.W. (1977) Effect of blue-green light on photosynthetic pigments and chloroplast structure in unicellular marine algae from six classes. *J. Phycol.*, **13**, 280–8.

VINCENT W.F. & GOLDMAN C.R. (1980) Evidence for algal heterotrophy in Lake Tahoe, California-Nevada. *Limnol. Oceanogr.*, **25**, 89–99.

YODER J.A. (1979) Effect of temperature on light-limited growth and chemical composition of *Skeletonema costatum* (Bacillariophyceae). *J. Phycol.*, **15**, 362–70.

4 Phytoplankton: population dynamics

The presence of a particular alga in a given body of water depends on a number of complex, spatial and temporal interactions among environmental factors affecting growth rate, behavioural characteristics of the alga and the activities of other organisms, both algal and non-algal. In the previous chapter the more immediate effects of major environmental factors on growth rate were discussed. In this chapter we broaden our view and consider aspects of metabolism, behaviour and competitive interactions which are superimposed on those environment–growth rate interactions. Competitive interactions with other algae are also important and are the subject of increasing experimental investigations. Planktonic blue-green algae are considered separately in this chapter because they possess several unique characteristics.

DIEL PERIODICITY

In spite of the ecological emphasis on the steady-state condition, most biological phenomena exhibit a periodicity of one form or another. Many aspects of phytoplankton physiology are known to flucutate over the 24 h (or diel) cycle; the amplitude of the diel change may be as great as the annual variation of certain parameters. It is still possible to consider many of these fluctuating systems as in steady state provided a suitable time frame is used. In natural phytoplankton communities diel periodicity has been observed in the following functions as summarized by Sournia (1974).

1. *Cell division.* The division of many species is phased (partially synchronized) in that a large percentage of the cells divide at a certain time of the day or night. If dividing or recently divided cells are easily recognized, growth rates of natural populations can be calculated by determining the percentage of dividing cells occurring in the population at intervals over the diel cycle (Weiler & Chisholm 1976). Physiologists take advantage of this response by manipulating the light–dark cycle to obtain highly synchronized cultures of algae for cell-cycle research. A number of factors, including nutrient status, temperature and photoperiod, may shift the time at which the peak in division occurs. Nocturnal division apparently is favoured by many species, especially those studied in culture, but more species need to be examined in the field.

2. *Buoyancy and vertical migration.* Several non-motile species are known to be more buoyant (i.e. they have a slower sinking rate) during the day than at night. In motile species there is a tendency for cells to accumulate at the surface or at intermediate depths during the day and to move downward in the evenings and often disperse during the night. Swimming speeds in the order of 0.5–2.0 m h^{-1} are strong enough to overcome some vertical currents. The migration patterns seem to be controlled by a phototactic response which follows a diel periodicity. Strong downward movement in some species at night also suggests the ability to respond to a gravitational field. The adaptive significance of the migrations might be related to avoiding predation and/or to finding higher concentrations of nutrients at depth. Vertical migration patterns in freshwater phytoflagellates from several different classes are illustrated in Fig. 4.1. *Oocystis*, a non-motile green alga, serves as a control in these samples because variations in its vertical distribution are largely the result of water turbulence.

3. *Chlorophyll a.* No general trends are evident in the many different diel chlorophyll curves which have been reported, although definite periodicity

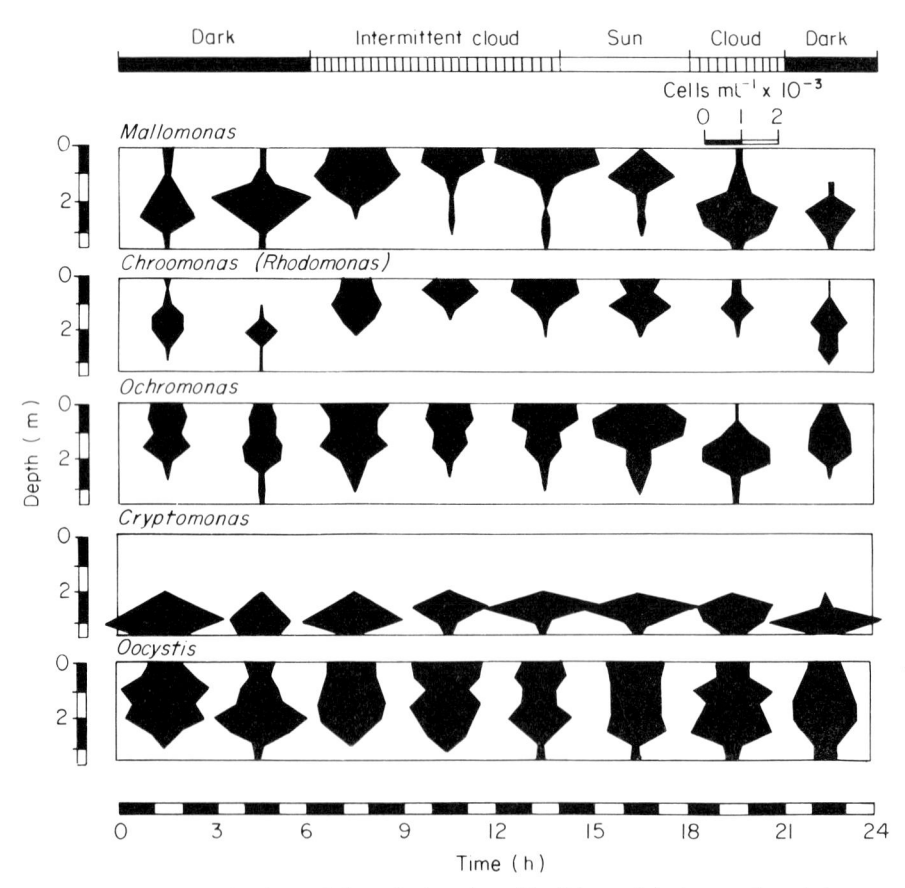

Fig. 4.1 Vertical distributions of phytoplankton in an English pond during a 24 h period. (After Happey-Wood 1976.)

may be observed for a given set of conditions, independent of grazing, cell division and vertical migration.

4. *Photosynthetic capacity.* Short-term, photosynthetic measurements at constant, saturating light intensities show maximum rates in the morning and minimum rates in the early evening. The response is independent of chlorophyll concentration, but the underlying mechanism is not understood. In a recent study of 24 species of marine phytoplankton in culture (Harding et al. 1981), 17 species were found to exhibit diel periodicity in P_{max}. The slope of the P versus I curve also fluctuated in synchrony with P_{max} in the four centric diatoms studied in more detail. Various patterns are observed in *in situ* photosynthesis measurements (at ambient light intensities), although they ordinarily track the available light with either morning or afternoon maxima.

5. *Nutrient uptake.* Nitrogen and phosphorus uptake rates are higher during the day than at night, a trend that reflects light stimulation of uptake rates. This response might partially explain why nutrient concentrations sometimes increase at night and decrease during the day in natural waters. Diel periodicity in $Si(OH)_4$ uptake reflects the phasing of cell division (wall formation) in diatoms. As shown in Fig. 4.2, there is significant variability

Fig. 4.2 Time-course of silicic acid incorporation observed in cultures of six species of marine planktonic diatoms grown on light–dark cycles. Data were normalized to the percentage of total incorporation to facilitate comparisons between species. *Lithodesmium undulatum* (a), *Thalassiosira fluviatilis* (b), *Chaetoceros gracilis* (c) and *Skeletonema costatum* (d) were grown on 12 : 12 light–dark cycles. *Ditylum brightwellii* (e) and *Thalassiosira pseudonana* (f) were grown on 14 : 10 light–dark cycles. (After Chisholm et al. 1978.)

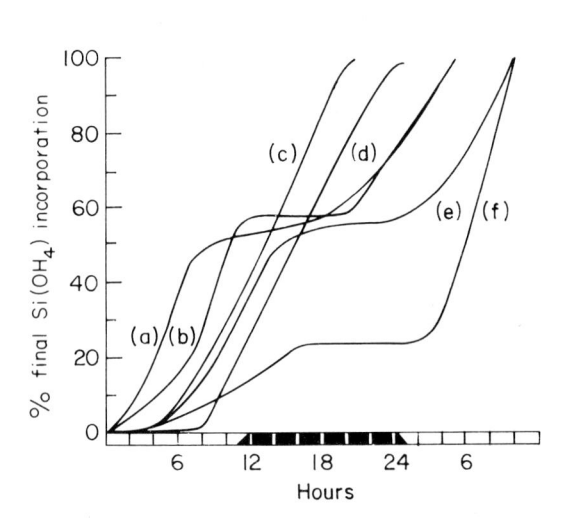

among species in the phasing of $Si(OH)_4$ incorporation. In natural populations (Fig. 4.3) the degree of incorporation periodicity appears to be inversely correlated with diatom species diversity (Chisholm et al. 1978). In the natural phytoplankton assemblage whose time course of silicic acid incorporation is shown in Fig. 4.3a, 95% of the diatoms present belonged to only two species. In the assemblage used to obtain the data in Fig. 4.3b, 95% of the diatoms present belonged to one of eight species.

6. *Bioluminescence.* The stimulated emission of light by certain dinoflagellates has a well-defined maximum at night. Among a number of theories

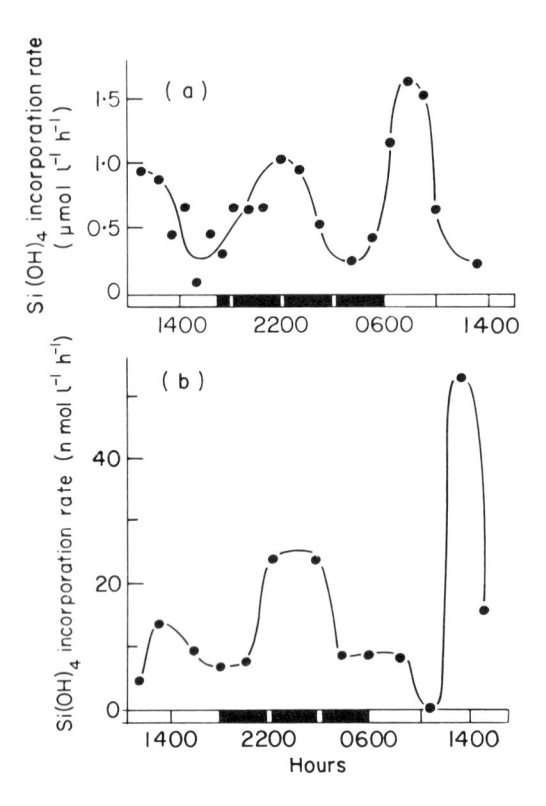

Fig. 4.3 Rates of silicic acid incorporation by natural phytoplankton populations incubated under natural light conditions. (a) Three diatom species were present in the sample from the Gulf of California, Mexico, on the 20 November 1974; (b) twelve species were present in the sample from La Jolla, California, on the 28 October 1975. (After Chisholm et al. 1978.)

on the adaptive significance of bioluminescence, it has been suggested recently that the light flash serves as an antipredation device (Porter & Porter 1979). Feeding on bioluminescent dinoflagellates by copepods produces a bright spot of light which makes the copepod more conspicuous to its predator. Thus copepods which feed on bioluminescent dinoflagellates would be at a selective disadvantage. The diel periodicity is readily explained because light flashes would be more visible at night. Copepod grazing is also more intense at this time.

A major concern of researchers investigating diel rhythms has been to decide if they are regulated by endogenous factors (i.e. circadian rhythms under the control of an internal biological clock) or exogenous factors (i.e. they are merely responding to diel fluctuations in the environment). Confirmation of endogenous control is obtained by demonstrating that the rhythm persists under constant environmental conditions. Persistent rhythms in axenic laboratory cultures have been reported many times, but it is more difficult to maintain a piece of nature under constant conditions. Several rhythms in natural phytoplankton populations anticipate the coming day or night rather than respond once the illumination has changed, providing some evidence for endogenous control in natural situations. From an ecological point of view, the question of endogenous versus exogenous control is not all that important. Under constant environmental conditions, circadian rhythms will revert to the free-running state, with a period of approximately, but seldom exactly, 24 hours. In natural situations, however, the free-running rhythm will be set to the 24 hour diel period by fluctuating environmental parameters, especially the light–dark cycle. In other words, the rhythm may be controlled by an endogenous clock, but that clock will be set every day to keep it running in phase with the 24 hour day. The biological clock regulates the potential response while the actual response will be determined largely by the environment. Remember also that periodicities in phytoplankton physiology cause diel fluctuations in the environment which will affect other organisms.

Despite the many advantages of steady-state chemostat cultures, they do not in fact truly mimic natural systems which undergo rhythmic changes. The imposition of a light–dark cycle on the chemostat (cyclostat culture) is a major improvement since the day–night cycle is probably the most significant cyclic environmental factor in most systems. To date, however, studies employing chemostats under light–dark cycles have not emphasized the diel changes in cell physiology. The available data are variable but follow the same trends listed above for natural populations. It is more difficult to test the effect of other environmental cycles, such as nutrient supply, O_2 and CO_2 concentrations, pH and grazing pressure in chemostat cultures, due to technical problems and because we do not yet have an adequate understanding of their diel periodicities.

It is interesting to speculate on the adaptive significance of diel rhythms in phytoplankton. By regulating the levels of various metabolic pathways during the diel period, the cell could be allocating its resources to take maximum advantage of the oscillating levels of environmental resources. Photosynthetic capacity, for example, is at a maximum some time during the daylight hours. It is also advantageous to have the biological rhythm in phase with the diel cycle rather than with the cell cycle since many species divide only once every two to four days in nature. It is easy for laboratory-bound phycologists to forget this important point since in many synchronized laboratory cultures all of the cells divide once a day. Hence we tend to equate diel timing with the developmental phases of the division cycle.

Perhaps the most puzzling aspect of diel rhythms is the diversity found among different species in the timing of maximum rates of activity; cell division is the best studied case. Given the fact the rhythms exist, one might expect that optimum-phase relationships would be similar for all species since they are all photoautotrophs with the same basic requirements. In addressing this problem Chisholm and co-workers (1978) cite references pointing out that phytoplankters could reduce interspecific competition by timing their maximum rates of various activities for different times of the day. For example, in an environment in which nutrients are more or less continually begin replenished via mineralization and zooplankton excretion, Species A could avoid competition with Species B by assimilating nutrients at a different time of day. The result would be niche separation on a temporal basis. Silicic acid incorporation by natural diatom populations (see

Fig. 4.3) is consistent with this hypothesis. It is also possible that diel rhythms in cell division are significant with respect to grazing mortality. Since zooplankton are known to select larger-sized particles within their prey size range and are also known to feed more intensively at certain times of the day, it would be to an alga's advantage to have completed cell division prior to the time of maximum feeding activity in order to present smaller cells to the predator. Testing these hypotheses concerning the adaptive significance of diel periodicity to phytoplankton and zooplankton promises to be an exciting area of research in the coming years.

SUSPENSION

A planktonic photoautotroph obviously must remain in the euphotic zone if it is to grow, although survival below the euphotic zone for various periods of time is not precluded. Smayda (1970) reviewed the extensive literature on phytoplankton suspension in the sea and concluded that water movements (e.g. Langmuir circulations; see Fig. 3.3) are more important to phytoplankton suspension than physiological and morphological adaptations on the part of the cells. In the absence of water movement, most non-motile phytoplankters will sink. Typical sinking rates for natural populations of phytoplankton range from 0.1 to 1–2 m d^{-1}. Therefore the question for most non-motile phytoplankton is not so much whether cells sink, but how fast they sink and how they modify their sinking rates. These adaptations along with examples of positive buoyancy (flotation) are discussed below.

Fat accumulation is perhaps the first physiological adaptation that comes to mind because lipid is the only class of organic molecules in the cell which is less dense than water. The amount of lipid in most cells, however, is not large enough to be an effective means of increasing cell buoyancy. The positively buoyant green alga *Botryococcus braunii* might be an exception to this generalization. The density of cytoplasm in freshwater and marine phytoplankton (1.01–1.03 and 1.03–1.10, respectively) is higher than the density of fresh water (< 1.0008) and sea-water (1.028) and cannot be reduced significantly to increase buoyancy.

The ionic composition of the vacuolar sap may be modified, however. In marine cells with large vacuoles, cell density may be lowered significantly by substituting lighter ions (e.g. NH_4^+, Na^+, Cl^-) for heavier ions (e.g. K^+, Ca^{++}, Mg^{++}, $SO_4^=$) in the vacuolar sap. Cells of the marine centric diatom, *Ditylum brightwelli*, grown on a light–dark cycle, undergo a 30–50% increase in sinking rate during the light period by accumulating K^+ and Cl^- and removing Na^+ from the vacuole (Anderson & Sweeney 1978). In the highly vacuolate, non-motile dinoflagellate, *Pyrocystis nóctiluca*, low concentrations of divalent ions in the vacuole are responsible for maintaining near neutral to positive buoyancy throughout the vegetative part of the cell cycle (Kahn & Swift 1978). Low light intensities act to increase buoyancy while low nutrient levels tend to decrease buoyancy. During reproduction, the cells become negatively buoyant because the vacuole is lost as the cytoplasm contracts. The net result of these buoyancy changes is to position the population toward the bottom of the euphotic zone. Ionic regulation is not effective in fresh water due to the low salinity.

Morphological features of cells also affect sinking rates. Cells with a higher surface area/volume ratio have more 'form resistance' and sink more slowly. Other things being equal (which they seldom are), a larger cell will sink faster than a smaller cell. Colony formation, usually considered an adaptation to increase buoyancy, actually results in increased sinking rates because of the decrease in the surface area/volume ratio. Various extensions of the cell surface (spines, etc.) commonly encountered in diatoms decrease sinking rates by increasing frictional drag, but this positive effect on buoyancy is offset in part by the addition of silica (density = 2.6) to the cell. Cell coverings of cellulose in dinoflagellates and calcium carbonate in coccolithophorids similarly may be considered mixed blessings, although it is possible that they and the surface extensions provide some protection from grazers.

Current hypotheses (Smayda 1970) suggest that chain formation and the diverse surface extensions characteristic of most planktonic diatoms function primarily as devices to cause turning, twisting and tumbling of the organism as it sinks. The advantage relates to nutrient absorption in oligotrophic waters. A cell which is stationary in the water soon depletes the nutrients in the microlayer near the cell, resulting in decreased nutrient uptake rates as the microlayer is replenished slowly by diffusion of nutrients from the surrounding water. As the cell sinks or rises in the water column, creating turbulence and renewing the water next to the cell, microgradients are broken down, the cell surface is exposed to higher nutrient concentrations and uptake rates are higher. Morphological features are arranged so as to cause a slow, zig-zag descent rather than a straight-line, vertical descent, thereby exposing the cell to a maximum amount of 'new' water with a minimum of vertical sinking in the water column. It is possible that the increased (two to four times) sinking rates commonly observed in nutrient-depleted cells are an adaptation to help the cell find more nutrient-rich water (Titman & Kilham 1976). Once more favourable nutrient conditions are encountered, the cells regain their former buoyancy and may be returned to surface waters by Langmuir circulations. This perspective of nutrient absorption points out another 'unnatural' aspect of culture studies; stirring reduces microgradients near the cell and results in erroneous (relative to the real world) estimates of nutrient uptake kinetics at low substrate concentrations. Motile cells, of course, can change their own position relative to the surrounding water (at the expense of metabolic energy) rather than depending on gravity to do it for them.

ALLELOPATHIC INTERACTIONS

Whereas the bulk of organic carbon excreted by phytoplankton (see Chapter 3) has no obvious function to either the excreting cell or to other algal cells, it is likely that algae also liberate small amounts of organic compounds which may have significant ecological consequences to the phytoplankton community. Some of the known cases involve the phenomenon of allelopathy which may be defined as a harmful effect of one alga on a second alga by means of a chemical compound liberated into the environment. Other inhibitors such as the antibacterial substances produced by many algae and the potent toxins produced by some bloom-forming dinoflagellates and blue-green algae will not be considered here. Allelopathic interactions usually are identified by testing the effect of the filtrate of a culture or water sample on the growth of a sensitive alga. Alternatively, the producer alga and sensitive alga are grown in cultures separated by a porous membrane through which small molecules can diffuse. It is very difficult to identify the specific compounds involved in these interactions due to their low concentrations.

Any extrapolation of allelopathic effects in cultures to natural situations must be done cautiously to ensure that: (1) the algae involved occur in the same habitat; (2) the population densities of the organisms and the concentrations of the allelopathic substances are representative of the natural situation; (3) other limiting factors are not involved; and (4) the possible effect of bacterial activity is not overlooked. These considerations were taken into account in a study of allelopathic interactions among blue-green algae forming a sequence of blooms in a eutrophic lake (Keating 1977). Seven species were isolated and tested for their effect on blue-green species which were dominant in earlier or later blooms. Culture filtrates of any given alga had an inhibitory or neutral effect on the growth of species immediately preceding it in the bloom sequence and had stimulating or neutral effects on species which immediately followed it in the bloom sequence. Similar results were obtained using filtrates of lake water obtained at the time various species were dominant. Interactions such as these may be one of the keys to understanding algal succession and should be investigated further.

RED TIDES

The earliest record of a red tide appears to be the reference in *Exodus* (7:20 and 21) to the Nile turning to blood and killing fish during the plagues of Egypt. We now know that red tides are massive blooms of dinoflagellates, usually belonging to the genera *Gonyaulax*, *Gymnodinium*, *Prorocentrum* or *Pyrodinium*; *Peridinium* blooms have also been observed in lakes. When cell densities exceed 500–1000 cells ml^{-1}, the water takes on the reddish tint which is responsible for the term red tide. They are most common in coastal and estuarine waters where the toxins produced by many of the species may cause fish kills or become concentrated in filter-feeding shellfish and cause paralytic shellfish poisoning in humans.

The mechanisms responsible for the development of red tides are not well understood. Attempts to correlate these sporadic and seasonal events with various chemical and physical parameters have been largely unsuccessful, especially in terms of predicting the occurrence of red tides. In recent years, however, good evidence has been obtained in support of two different explanations for the initiation of red tides.

Along the New England and eastern Canadian coastline, outbreaks of the toxic dinoflagellate, *Gonyaulax tamarensis*, seem to be initiated by germination of thick-walled, ovoid cysts called hypnozygotes. On Cape Cod there is a direct correlation between the presence of these cysts in the sediment and the occurrence of red tides the preceding and following springs. Hypnozygote formation is a sexual process triggered by unknown environmental or biological factors. Maturation and germination of hypnozygotes are temperature-dependent processes (Anderson 1980). Hypnozygotes stored at 22°C mature within a month and germinate in response to a *decrease* in temperature, thereby contributing to a fall bloom. Hypnozygotes stored at 5°C require several months to mature, germinate in response to *increasing* temperature and therefore function as an overwintering phase of the life cycle that initiates the spring bloom. Of course the development and intensity of the red tide are a function of vegetative growth and are therefore dependent on many parameters in addition to temperature.

Other explanations must be sought for the late spring and summer red tides caused by the non-toxic dinoflagellate, *Prorocentrum mariae-lebouriae*, in the northern end of Chesapeake Bay (USA) (Fig. 4.4). No *Prorocentrum* cysts are found in the sediments and the vegetative cells do not persist in the northern bay during the winter. They cannot grow fast enough at the combination of low salinities and low temperatures to overcome the wash-out effect of the rivers entering the northern bay. Research by Tyler & Seliger (1981) has shown that the blooms are initiated by vegetative cells which have been transported by subsurface currents from the southern end of the bay, a distance of about 240 km. During the winter months *Prorocentrum* washed from the northern bay is found growing at the higher salinities near the mouth of the bay. The cells become concentrated at convergence zones (A in Fig. 4.4) between bay water and water entering the bay from the James and York Rivers. In those years when heavy, winter rains occur over the watershed of the northern bay, this lens of concentrated *Prorocentrum* cells is forced beneath the southward-flowing, freshwater run-off. The lens is subsequently transported north with northward-flowing, subsurface currents until the bottom topography of the bay forces the water to the surface (B in Fig. 4.4). When this nutrient-rich, bottom water mixes with the warm, illuminated, surface waters, the *Prorocentrum* population rapidly grows to bloom proportions.

Now that the inoculation mechanisms are understood, these red tide occurrences are readily predicted from winter–spring river flow in the southern and northern regions of the bay. Red tides do not occur unless the proper sequence of events has taken place. Given the proper conditions, subsurface concentrations of *Prorocentrum* in water of higher salinites may

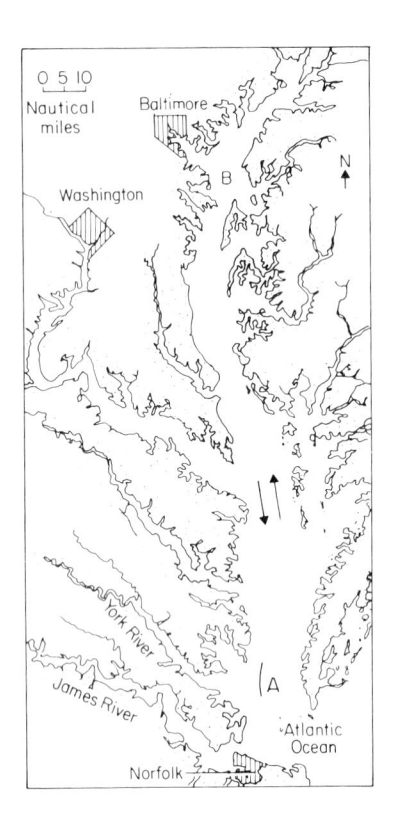

Fig. 4.4 Chesapeake Bay, USA. A indicates the zone of convergence where *Prorocentrum* cells are concentrated in the winter. B indicates the area of red tides in the late spring and summer months. The arrows denote southward-flowing, less saline, surface water and northward-flowing, more saline, subsurface water containing *Prorocentrum* cells. (After Tyler & Seliger 1981.)

be sought in the southern bay to confirm that another red tide can be expected in the northern bay within the next few months. Scattered reports in the literature suggest that concentration and subsurface transport phenomena might be useful approaches to explore in attempts to explain red tides in other localities.

Although the subsurface, northward transport of *Prorocentrum mariae-lebouriae* in the Chesapeake Bay is largely the result of physical processes, the physiology of this dinoflagellate is well adapted to the transport process. During the trip north the cells retain their viability and photosynthetic capacity at light intensities between 0.1 and 1% of surface values. The cells undergo typical shade adaptation (increased chlorophyll per cell and lower respiration rates) with the result that the compensation point for photosynthesis is within the range of the *in situ* illumination. One might expect that because of positive phototaxis on the part of *Prorocentrum*, they could migrate into the surface waters. That they are retained in the northward-flowing, bottom water is apparently the result of a loss of motility experienced when crossing the halocline into less saline, surface water. Once delivered to the northern end of the bay, the bloom often persists into the summer months because the cells migrate to deeper, nutrient-rich water at night. This diel, vertical migration pattern was not observed earlier in the season before nutrients became depleted in the surface waters.

BLUE-GREEN ALGAE

Anoxygenic photosynthesis

Blue-green algae normally photosynthesize by the typical higher plant process in which two photosystems are involved—water provides the reductant and O_2 is evolved (oxygenic). Bacterial photosynthesis is very different; only photosystem I is involved—sulphide, sulphur, thiosulphate or organic compounds provide the reductant and O_2 is not evolved (anoxygenic). Under certain conditions some blue-green algae are capable of switching over to anoxygenic photosynthesis, presumably having retained this capacity during their evolution from photosynthetic bacteria. In *Oscillatoria limnetica* anoxygenic photosynthesis is induced within two hours in the light in the presence of H_2S (Padan 1979). Growth rates under anoxygenic photosynthesis are similar to those under oxygenic photosynthesis. When O_2 is reintroduced into an induced culture, oxygenic photosynthesis begins immediately and the anoxygenic capacity is diluted out by growth.

Eukaryotic algae and photosynthetic bacteria are restricted to environments that are photoaerobic and photoanaerobic respectively. The selective advantage of facultative anoxygenic photosynthesis is that cells with this capacity are well adapted to environments that periodically fluctuate between photoaerobic and photoanaerobic conditions (Padan 1979). Solar Lake (Sinai) is a well-studied example of such a habitat. During the period of water column stagnation the hypolimnion becomes anaerobic with a high concentration of H_2S. *Oscillatoria limnetica*, growing as a facultative anoxygenic phototroph, forms dense blooms under these conditions. During the brief period of spring overturn, the lake becomes aerobic and *O. limnetica* grows throughout the water column as an oxygenic phototroph alongside the eukaryotic phytoplankton.

Nitrogen fixation

The nitrogen metabolism of blue-green algae is unique among the algae in at least two ways: they are able to store organic nitrogen and to fix N_2. Reduced nitrogen is stored as a copolymer of aspartic acid and arginine in a 1 : 1 molar ratio. The compound is localized in small bodies termed cyanophycin granules (also called structured granules) which accumulate during nitrogen-sufficient growth and in akinetes. No comparable N-storage compounds have been reported in other classes of algae. The biliprotein pigments in blue-green algae also act as a nitrogen reserve in that the cellular levels of these compounds decrease rapidly upon nitrogen depletion.

N_2 fixation is essential to the earth's nitrogen budget since it replaces fixed nitrogen lost in denitrification. The process is limited to prokaryotes and therefore to the blue-greens among the algae. Since the nitrogen-fixing enzyme, nitrogenase, is irreversibly inactivated by O_2, it must be protected from O_2 by some mechanism. The problem is compounded in O_2-evolving

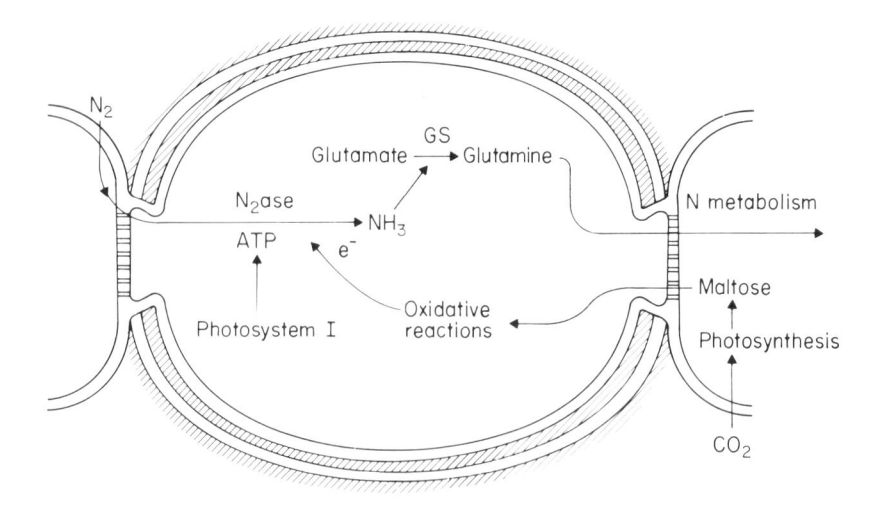

Fig. 4.5 Nitrogen and carbon flow associated with N_2 fixation in the heterocysts of the blue-green algae. N_2ase = the nitrogenase enzyme and GS = glutamine synthetase; oxidative reactions include the oxidative pentose phosphate pathway. Polar nodules are not shown. (After Haselkorn 1978.)

photoautotrophs such as blue-green algae. In most filamentous blue-greens that fix N_2, the heterocyst is the answer to these problems. In the absence of combined forms of nitrogen and under poorly understood controls by neighbouring cells, a vegetative cell will cease dividing and begin differentiating into a larger, rounded, pale cell with a thickened cell wall (Fig. 4.5) and distinctive polar nodules at either end. In addition to its thicker wall, which probably functions to exclude oxygen, the heterocyst is modified for N_2 fixation by the loss of several components: (1) the O_2-evolving photosystem II; (2) the biliprotein pigments associated with photosystem II (hence the pale colour of heterocysts); and (3) ribulose 1,5-bisphosphate carboxylase, the CO_2-fixing enzyme of photosynthesis (Stewart 1980). ATP for N_2 fixation is provided through cyclic photophosphorylation in photosystem I, while reductant for N_2 fixation is obtained through catabolism of reduced carbon (mostly maltose) which is imported from adjacent vegetative cells. Ammonia, the product of N_2 fixation, is exported from the cell as glutamine which is then further metabolized in vegetative cells. Bacteria are often observed attached to heterocysts of planktonic blue-green algae (*Anabaena, Aphanizomenon*), especially in blooms. Evidence has been presented that this relationship is mutualistic with the algae providing organic substrates for the bacteria and the bacteria reducing O_2 tensions in microzones near the heterocyst to facilitate N_2 fixation in highly oxygenated, surface waters (Paerl & Kellar 1978).

Some non-heterocyst-producing blue-greens can also fix N_2, but only at much reduced O_2 tensions, environments not normally encountered in the euphotic zone in open water. A planktonic *Oscillatoria* (*Trichodesmium*) found in tropical and subtropical seas displays another non-heterocystous mechanism for protecting nitrogenase from O_2 inactivation. Filaments which pass through the centre of the colony differentiate a few lightly pigmented cells which lack the capacity for CO_2 fixation and presumably do not evolve O_2. N_2 fixation rates drop rapidly if these loosely organized colonies are dispersed by shaking in oxygenated water, but do not drop if dissolved O_2 is removed before dispersing the colonies. These observations suggest that the pale cells in the centre of the colony are the site of N_2 fixation and explain why blooms of this organism are always found in calm seas (Carpenter & Price 1976). A more recent study (Bryceson & Fay 1981) has confirmed and extended these observations, but conclusive evidence for N_2 fixation by this alga requires axenic cultures, a goal that has not yet been achieved.

Gas vacuoles

Planktonic blue-green algae possess flotation devices called gas vacuoles which are unique to prokaryotic organisms (Walsby 1977). A single gas vacuole consists of a number of closely packed cylinders with tapered ends (gas vesicles), each 70 nm wide and 200–1000 nm long and limited by a 'membrane' of pure protein. Gas vesicles are strong but brittle structures which collapse in an 'all-or-nothing' manner under sufficient pressure. They appear to be synthesized by a self-assembly process; the gaseous contents are in equilibrium with the surrounding water. Planktonic blue-green algae are able to position themselves in the water column by regulating the number of gas vesicles which in turn affects cell buoyancy. At high light intensities, cytoplasmic turgor pressure increases sufficiently to collapse some of the weaker gas vesicles, increasing cell density and causing the cells to sink. This increase in turgor pressure is the result of light-stimulated K^+ uptake and of increased concentrations of osmotically active, photosynthetic products (Allison & Walsby 1981). At lower light intensities, turgor pressure declines, gas vesicles (forming continually through self-assembly) increase in number and the alga regains its positive buoyancy. The role of gas vacuoles in regulating buoyancy is illustrated by the three examples which follow.

Planktonic *Oscillatoria* species often form discrete layers near or below the thermocline in lakes during periods of summer stratification. Displacement experiments (Fig. 4.6) show that *O. agardhii* regulates its buoyancy in an attempt to return to its original position in the water column. This response persisted at constant temperature suggesting that light intensity rather than temperature was the controlling parameter. Another experiment showed that the preference for deeper water is also related to nutrient availability. When an enclosed water column was enriched with nutrients (Fig. 4.6), *Oscillatoria* was found in large numbers at the surface. The effect of nutrients on buoyancy might be explained as follows. Under nutrient-sufficient conditions, photosynthetic products are more rapidly incorporated

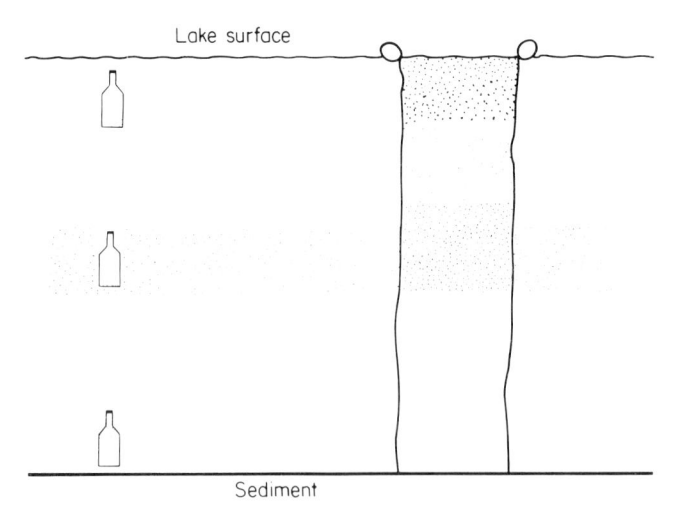

Fig. 4.6 Buoyancy control in *Oscillatoria agardhii*. Natural populations form discrete layers at intermediate depths in stratified lakes in the summer. Left, the results of a displacement experiment showing the distribution of filaments in bottles after one day. Right, a clear polyethylene tube was inserted into the water column with a minimum of disturbance and enriched with nutrients for one month. (After Klemer 1976.)

into macromolecules associated with growth and therefore higher light intensities are required before enough osmotically active products accumulate to initiate gas-vesicle collapse. It is also possible that improved nitrogen status would result in an increased cellular content of gas-vesicle protein with concomitant increases in the number of gas vesicles and buoyancy.

Other planktonic blue-green algae such as *Anabaena*, *Aphanizomenon* and *Microcystis* tend to occupy the epilimnion of lakes. In eutrophic Lake Mendota, (Wisconsin, USA), wind stress frequently mixes the epilimnion to a depth of 10–12 m, well below the euphotic zone which only extends to about 4 m in the summer due to blooms of blue-green algae. During calm periods the algae rapidly float towards the surface (Konopka *et al.* 1978). Buoyancy is provided by gas vesicles which have accumulated during the average, low light intensities experienced during mixing. During extended calm periods *Aphanizomenon flos-aquae* occasionally retreated from the surface during the afternoon, due to increased turgor pressure which collapsed some gas vesicles. For hydrodynamic reasons the large colony size of these algae allows them to migrate quickly, in contrast to the individual filaments of *Oscillatoria* which change vertical position slowly. Thus colony-forming species are adapted to turbulent water masses in which they must return to the euphotic zone during calm weather, whereas species with isolated filaments are adapted to stable water masses in which they remain in well-defined, subsurface layers.

The marine, planktonic, blue-green, *Oscillatoria* (*Trichodesmium*) would sink rapidly due to its large colony size (a requirement for N_2 fixation; see

p. 65) if it did not contain gas vacuoles. The vesicles in this alga are two to six times stronger than those in freshwater forms, an attribute which precludes the regulation of buoyancy through changes in osmotic pressure but also prevents vesicle collapse (and loss of buoyancy) when cells are carried to depths of over 100 m by large-scale, oceanic Langmuir circulations. These slow-growing algae therefore float back to the surface following a period of stormy weather.

Blooms

Algal blooms in lakes are usually composed of colony-forming, unicellular and filamentous blue-green algae, often belonging to one of the following genera: *Microcystis, Anabaena, Aphanizomenon,* or *Gloeotrichia* (Reynolds & Walsby 1975). Except for the presence of gas vacuoles, bloom-forming species do not appear to have any special physiological characteristics beyond the general tendency of blue-green algae to favour mildly alkaline, warmer and slightly eutrophic waters. Excessively eutrophic or polluted habitats are dominated by small green algae and euglenoids which form blooms, but not in the sense of the surface phenomena associated with blue-greens. Blooms of blue-green algae occur when populations which have been growing unnoticed below the surface for several weeks, suddenly float to the surface during periods of calm weather, as discussed above. The algae may remain trapped at the surface by their excess buoyancy if they are incapable of increasing turgor pressure to collapse some gas vesicles. Decreased rates of photosynthesis caused by nutrient and/or CO_2 deficiency and low levels of K^+ in the surface waters are possible contributing factors. Intense sunlight at the surface causes further senescence and death of the algae. A light wind can concentrate the surface bloom along a shore line as a smelly scum which lowers water quality and has killed livestock and waterfowl that drink the water. Several different toxins, including alkaloids and cyclic polypeptides, have been identified in different bloom-forming species. Blue-green algal blooms therefore seem to be incidental events which occur when a period of calm weather happens to coincide with excess and uncontrollable buoyancy in a substantial population of cells. To the extent that human activity moderately enriches the nutrient levels of lakes, we can expect blooms to occur with increasing frequency.

RESTING CELLS

Many phytoplankton species disappear from the water column for extended periods of time during the year and mechanisms for survival during these periods are not well understood. It is possible that the organism is present but at such low concentrations that it is undetectable (e.g. 1 cell m^{-3}). These 'seed' organisms would serve as an inoculum for an increase in the population upon the return of favourable conditions. It is also possible that some cells survive in darkness below the euphotic zone until currents return them to the surface. A number of planktonic algae can survive periods of darkness of several months or even years (Antia 1976). Survival is both species and temperature dependent. Cold-water forms survive longer at lower temperatures while the opposite is true of species characteristic of warm waters. In nature, heterotrophic activity (p. 49) might help provide for maintenance metabolism during extended periods of darkness.

Various types of sexual and asexual resting spores produced by phytoplankton appear to be specific adaptations designed to survive unfavourable conditions. The zygospore of volvocalean green algae survives storage in dry soil for many years. Chrysophytes typically produce a siliceous statospore, but its resistance to adverse conditions has not been tested. The hypnozygote of dinoflagellates was discussed earlier (p. 61). Akinetes produced by blue-green algae are well known for their longevity (100 years or more). Recent research on planktonic blue-green algae, however, has cast doubt on the significance of akinetes as an overwintering stage. Sporulation is not always observed and akinetes may germinate soon after being formed, a non-adaptive strategy for an overwintering structure. Although akinete formation has been induced in laboratory cultures of planktonic species, the specific physical or chemical factors which induce it are still unknown (Rother & Fay 1979). In centric diatoms it is the asexual resting cell, rather than the sexual auxospore, that apparently enables the alga to survive unfavourable conditions. Nitrogen limitation has been identified as the factor triggering resting-cell formation and light appears to be the most important factor initiating subsequent germination (Hollibaugh et al. 1981). Resting cells therefore must be returned to the euphotic zone before they will contribute to renewed vegetative growth of the species.

COMPETITION AND THE PARADOX OF THE PLANKTON

The fundamental ecological principle of competitive exclusion predicts that relatively homogeneous environments such as the surface, mixed layers of lakes and oceans should contain very few species which have similar ecological requirements. Since all species of planktonic algae are essentially photoautotrophs with similar needs which must be fulfilled from their mutually held surroundings, competition for resources, especially nutrients, should result in the elimination of all but the one species which is best able to utilize the limiting resources. The observation that 10-50 phytoplankton species commonly coexist in the same body of water has been termed the 'paradox of the plankton' by Hutchinson (1961). How can so many different algae use the same body of water in very similar ways?

One possible explanation is that species-specific, nutrient-utilization characteristics result in different species each being limited simultaneously by a different nutrient. Direct competition is avoided and potentially as many species can coexist as there are limiting nutrients. In an experimental test of this hypothesis (Tilman 1977), two, freshwater, planktonic diatoms

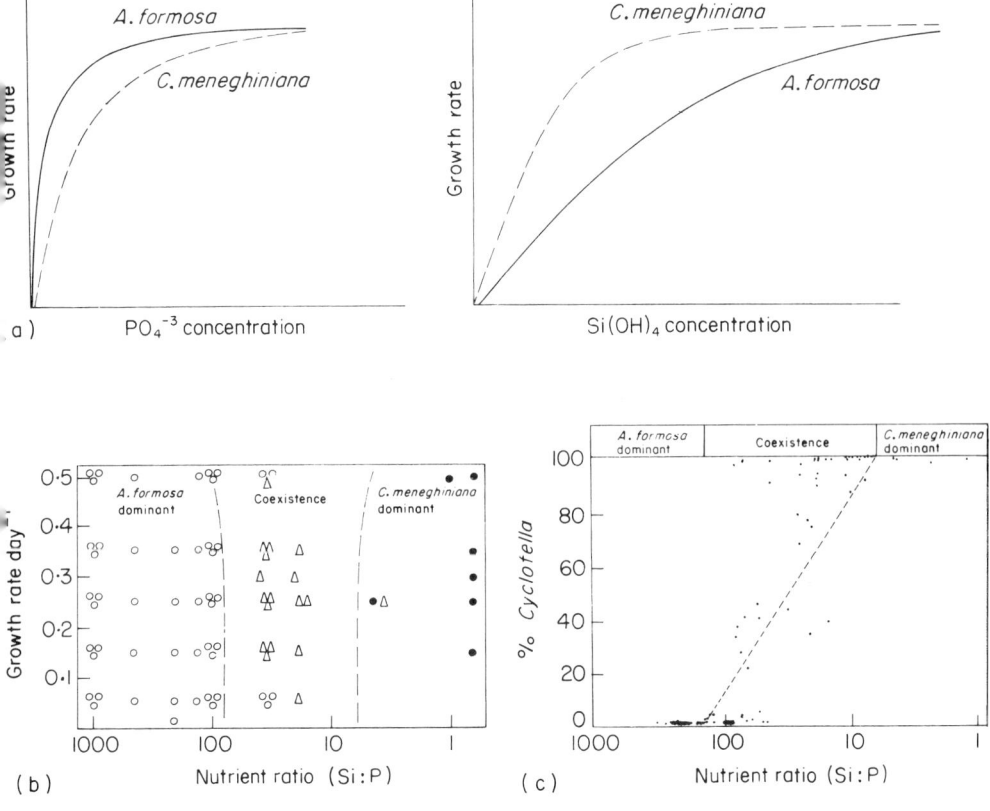

Fig. 4.7 Competition between *Asterionella formosa* and *Cyclotella meneghiniana* along a resource gradient. (a) The nutrient-related growth kinetics of each species in axenic culture. (b) The results of competition experiments as a function of the growth rate and Si : P ratio; *Asterionella* as dominant (open circle); *Cyclotella* as dominant (solid circle); and both species coexisting (open triangle). (c) The relative abundance of *Cyclotella* as a percentage of the total number of *Asterionella* and *Cyclotella* cells in samples from Lake Michigan plotted against the Si : P ratios measured in these same samples. (Data from Tilman 1977; adapted from Kilham & Kilham 1980.)

with similar maximum growth rates were grown together in semi-continuous culture (diluted once each day) at different Si : P ratios. As shown diagrammatically in Fig. 4.7a, *Asterionella formosa* has a lower half saturation constant for phosphate-limited growth than *Cyclotella meneghiniana*, whereas it has the higher K_s for silicate-limited growth. Thus at low P concentrations (high Si : P ratios) *Asterionella* should have the competitive advantage while at low Si concentrations (low Si : P ratios), *Cyclotella* should have the competitive advantage. At intermediate Si : P ratios the two species should stably coexist because each is limited by a different nutrient, Si for *Asterionella* and P for *Cyclotella*. The predicted results were obtained when these two diatoms were grown together at various growth rates at different Si : P ratios (Fig. 4.7b). The ecological significance of this laboratory study was confirmed by field observations of the relative abundance of these two species in Lake Michigan as a function of ambient Si : P ratios (Fig. 4.7c).

Grazing activity by zooplankton might also improve the chances for coexistence among species of planktonic algae. By reducing the total algal biomass, grazing will reduce demand on nutrient resources and therefore reduce the severity of competitive interactions. More direct effects also may be involved because many zooplankton species are known to graze selectively on the basis of prey size. By selective grazing on an abundant size class (the superior competitors under prevailing conditions), zooplankton would be contributing to the survival of inferior competitors. Zooplankton activity will not always increase phytoplankton diversity however (Porter 1977). Intense grazing could eliminate some species from the water column. When the dominate algae are large, inedible, colonial or filamentous species, even moderate grazing on the smaller edible species could reduce diversity.

Other explanations of the paradox stress that the mixed layer of lakes and oceans is not as homogeneous as is generally assumed. Well-known vertical gradients in light, nutrients and temperature during calm weather provide spatial heterogeneity allowing localized increases in those populations which find themselves in favourable conditions. Organisms which can seek out optimum points in these gradients through motility or buoyancy control can also minimize direct competition with other species. Diel rhythms in physiological functions (p. 57) are likely to be significant factors in providing temporal heterogeneity in resource utilization. The environment also is changing on a more extended time scale, providing favourable conditions for some species for a period of days or weeks and then shifting in favour of another species. Observations of natural assemblages which suggest the stable coexistence of many species, in fact, might represent a transition state in which some are declining in abundance and others are increasing. In summary, it would appear that phytoplankton diversity can be explained by a number of different mechanisms which probably function concurrently; the challenge is to provide experimental confirmation.

SEASONAL SUCCESSION

The annual changes in phytoplankton abundance and community structure which occur with some degree of predictability in temperate waters are probably the best-studied aspects of algal ecology. The subject is covered in some detail in many other publications and will only be outlined here as a means of summarizing and integrating the information which has been presented in these two chapters.

During the winter season, with low temperature, low light intensities and short days, phytoplankton biomass and productivity are generally low in spite of elevated nutrient concentrations. The phytoplankton is dominated by cold-tolerant species of several algal classes which on occasion may account for a surprising amount of production even under an ice cover. Increasing light (higher intensities and longer days) in the early spring is thought to be the principal factor which stimulates the spring outburst of phytoplankton production since temperature often remains low during this period. Recall that light-limited photosynthesis is independent of temperature (p. 26). In deeper waters the spring increase may be delayed if mixing is sufficiently deep to take the cells out of the euphotic zone often enough to keep the 'average' light intensity at limiting levels. As the water warms and stratification begins to set in, the mixed layer becomes shallower and the algae are able to take full advantage of the nutrient stockpile. The smaller, more edible, rapidly growing species often dominate the spring bloom because of the lag in the appearance of slower-growing zooplankton. The bloom continues until nutrient depletion and increased predation signal the summer decline in phytoplankton production. In both freshwater and marine systems, diatoms are usually the dominant species of the spring blooms and are succeeded by dinoflagellates and other motile species which dominate the less abundant, but more diverse, summer flora. Green algae are also common in the spring and summer phytoplankton of temperate lakes and may be joined by blue-green algae in more eutrophic situations. Larger, less edible species tend to be more common during the summer. A smaller autumn outburst of diatoms is sometimes found during the overturn that occurs with the end of summer stratification.

More detailed explanations for seasonal succession have been suggested in addition to the general involvement of light, nutrients, temperature and grazing pressure mentioned above. It is likely that the decline in diatom abundance is related to the slow regeneration of silicate in the mixed layer. Motile cells are possibly at an advantage during the nutrient-deficient, stratified periods because they can seek out favourable positions in the water column during calm weather. At the metabolic level, the overall summer decline in phytoplankton abundance may be explained in part, by a decrease in the photosynthesis/respiration ratio. Nutrient deficiency lowers photosynthetic rates while increasing temperatures raise respiration rates and the result is less net production (Devol & Packard 1978).

The preceding overview, while satisfying in a broad sense, does not begin to explain the extensive changes in species abundance which occur within this annual cycle. Ultimately we are obliged to account for the sequential waxing and waning and/or coexistence of individual species within the spring outburst. The key to the success of any given species is likely to be unique to that species and will depend on its physiological capabilities and tolerances as well as on the influence of other organisms preceding it and coexisting with it in the successional sequence. It is likely that a different explanation will be applicable at another time and place. As reviewed in these chapters, considerable progress is being made and we can look forward to new and exciting advances in this field.

EVOLUTIONARY ECOLOGY

Since survival in the evolutionary sense is clearly more than simply a case of who can grow the fastest, we must develop a greater understanding of the various strategies which have evolved among the algae. Kilham and Kilham (1980) have considered the selective pressures acting upon phytoplankton in terms of the concepts of *r*- and *K*-selection, currently in use for higher organisms. Their ideas, many of which are admittedly speculative, are summarized below in the hope of stimulating further research into the adaptive significance of the biological properties of phytoplankton.

r-selected species are opportunistic (weedy) species that have evolved in fluctuating environments. They are able to take advantage of temporary but favourable circumstances in their environment by virtue of their rapid growth rates. They are adapted to environments in which: (1) resources are plentiful; (2) population densities are below carrying capacity; and (3) competitive interactions are at a minimum. Among phytoplankton, *r*-selected species tend to be small cells (when compared with other species in the same taxonomic group) with high maximum growth rates. They tend to be characteristic of eutrophic habitats with high productivity (e.g. estuaries) and to be found at the earlier (spring) stages of seasonal succession in temperate waters. Diatoms as a group appear to have characteristics of *r*-selected species.

K-selected species are at the other extreme of this continuum. They have evolved in environments that are relatively stable over time. In these environments the supply of resources is matched by the demand for resources which means that population densities are at or close to carrying capacity and that competitive interactions are more severe. *K*-selected species are more efficient at resource utilization, tending to produce fewer offspring which have greater competitive abilities. Among phytoplankton, *K*-selected species tend to be the larger species within a given taxonomic group and to have slower, maximum growth rates. They tend to be better competitors with adaptations such as lower K_s values for nutrient uptake and the production of allelopathic substances. *K*-selected phytoplankton species are characteristic of oligotrophic habits with low productivity (e.g. the central gyres of the oceans) and are found during the later (summer) stages of seasonal succession in temperate waters. Dinoflagellates as a group appear to have characteristics of *K*-selected species.

REFERENCES

ALLISON E.M. & WALSBY A.E. (1981) The role of potassium in the control of turgor pressure in a gas-vacuolate blue-green alga. *J.exp.Bot.*, **32**, 241–9.

ANDERSON D.M. (1980) Effects of temperature conditioning on development and germination of *Gonyaulax tamarensis* (Dinophyceae) hypnozygotes. *J. Phycol*, **16**, 166–72.

ANDERSON L.W.J. & SWEENEY B.M. (1978) Role of inorganic ions in controlling sedimentation rate of a marine centric diatom *Ditylum brightwelli*. *J. Phycol.*, **14**, 204–14.

ANTIA N.J. (1976) Effects of temperature on the darkness survival of marine phytoplanktonic algae. *Microbial Ecol.*, **3**, 41–54.

BRYCESON I. & FAY P. (1981) Nitrogen fixation in *Oscillatoria* (*Trichodesmium*) *erythraea* in relation to bundle formation and trichome differentiation. *Mar. Biol.*, **61**, 159–66.

CARPENTER E.J. & PRICE C.C. IV (1976) Marine *Oscillatoria* (*Trichodesmium*): Explanation for aerobic nitrogen fixation without heterocysts. *Science*, **191**, 1278–80.

CHISHOLM S.W., AZAM F. & EPPLEY R.W. (1978) Silicic acid incorporation in marine diatoms on light : dark cycles: Use as an assay for phased cell division. *Limnol. Oceanogr.*, **23**, 518–29.

DEVOL A.H. & PACKARD T.T. (1978) Seasonal changes in respiratory enzyme activity and productivity in Lake Washington microplankton. *Limnol. Oceanogr.*, **23**, 104–11.

HAPPEY-WOOD C.M. (1976) Vertical migration patterns in phytoplankton of mixed species composition. *Br. phycol. J.*, **11**, 355–69.

HARDING L.W. Jr., MEESON B.W., PRÉZELIN B.B. & SWEENEY B.M. (1981) Diel periodicity of photosynthesis in marine phytoplankton. *Mar. Biol.*, **61**, 95–105.

HASELKORN R. (1978) Heterocysts. *Ann. Rev. plant Physiol.*, **29**, 319–44.

HOLLIBAUGH J.T., SEIBERT D.L.R. & THOMAS W.H. (1981) Observations on the survival and germination of resting spores of three *Chaetoceros* (Bacillariophyceae) species. *J. Phycol.*, **17**, 1–9.

HUTCHINSON G.E. (1961) The paradox of the plankton. *Am. Natural.*, **95**, 137–45.

KAHN N. & SWIFT E. (1978) Positive buoyancy through ionic control in the non-motile marine dinoflagellate *Pyrocystis noctiluca* Murray ex Schuett. *Limnol. Oceanogr.*, **23**, 649–58.

KEATING K.I. (1977) Allelopathic influence on blue-green bloom sequence in a eutrophic lake. *Science*, **196**, 885–7.

KILHAM P. & KILHAM S.S. (1980) The evolutionary ecology of phytoplankton. In Morris I. (ed.), *The Physiological Ecology of Phytoplankton*, pp. 571–97. Blackwell Scientific Publications, Oxford.

KLEMER A.R. (1976) The vertical distribution of *Oscillatoria agardhii* var *isothrix*. *Arch. Hydrobiol.*, **78**, 343–62.

KONOPKA A., BROCK T.D. & WALSBY A.E. (1978) Buoyancy regulation by planktonic blue-green algae in Lake Mendota, Wisconsin. *Arch. Hydrobiol.*, **83**, 524–37.

PADAN E. (1979) Facultative anoxygenic photosynthesis in cyanobacteria. *Ann. Rev. plant Physiol.*, **30**, 27–40.

PAERL H.W. & KELLAR P.E. (1978) Significance of bacterial–*Anabaena* (Cyanophyceae) associations with respect to N_2 fixation in freshwater. *J. Phycol.*, **14**, 254–60.

PORTER K.G. (1977) The plant–animal interface in freshwater ecosystems. *Am. Sci.*, **65**, 159–70.

PORTER K.G. & PORTER J.W. (1979) Bioluminescence in marine plankton: A coevolved antipredation system. *Am. Natural.*, **114**, 458–61.

REYNOLDS C.S. & WALSBY A.E. (1975) Water-blooms. *Biol. Rev.*, **50**, 437–81.

ROTHER J.A. & FAY P. (1979) Blue-green algal growth and sporulation in response to simulated surface bloom conditions. *Br. phycol. J.*, **14**, 59–68.

SMAYDA T.J. (1970) The suspension and sinking of phytoplankton in the sea. *Oceanogr. mar. Biol. Ann. Rev.*, **8**, 353–414.

SOURNIA A. (1974) Circadian periodicities in natural populations of marine phytoplankton. *Adv. mar. Biol.*, **12**, 325–89.

STEWART W.D.P. (1980) Some aspects of structure and function in N_2-fixing cyanobacteria. *Ann. Rev. Microbiol.*, **34**, 497–536.

TILMAN D. (1977) Resource competition between planktonic algae: An experimental and theoretical approach. *Ecology*, **58**, 338–48.

TITMAN D. & KILHAM P. (1976) Sinking in freshwater phytoplankton: Some ecological implications of cell nutrient status and physical mixing processes. *Limnol. Oceanogr.*, **21**, 409–17.

TYLER M.A. & SELIGER H.H. (1981) Selection for a red tide organism: Physiological responses to the physical environment. *Limnol. Oceanogr.*, **26**, 310–24.

WALSBY A.E. (1977) The gas vacuoles of blue-green algae. *Sci. Am.*, **237**(2), 90–7.

WEILER C.S. & CHISHOLM S.W. (1976) Phased cell division in natural populations of marine dinoflagellates from shipboard cultures. *J. exp. mar. Biol. Ecol.*, **25**, 239–47.

5 Benthic algae

The algae considered in this chapter include all those aquatic forms living in or on a substratum of some sort, exclusive of the seaweeds which are considered in Chapter 6. Pennate diatoms, green algae and blue-green algae are the most common forms (Fig. 5.1), although euglenoids, chrysophytes and xanthophytes also contribute to this very rich algal flora. The high species diversity is due to the tremendous variety and heterogeneity of the microhabitats which occur in shallow water. Algae are found growing on rocks, sandy and muddy sediments, animals, aquatic vascular plants and other algae. Algae which are not actually attached to a substratum but are found loosely associated with the attached algae, vascular plants and debris in shallow water are sometimes referred to as tychoplankton or metaphyton. A number of recent reviews have included sections on benthic algae (Hutchinson 1975; Wetzel 1975; Whitton 1975; McIntire & Moore 1977; Patrick 1977).

The environment of streams and rivers and the littoral zone of lakes and oceans, is different in several respects from the open-water environment. The nearby sediments with their abundant populations of heterotrophs contribute to higher concentrations of dissolved nutrients than are found in open-water habitats. Attached algae have the additional advantage of having nutrients delivered to their doorstep by currents whereas planktonic species must sink or otherwise change their position in the water column to renew that microlayer of water next to the cell from which nutrients are recovered. Despite the fact that benthic algae are in no danger of sinking out of the euphotic zone, they are often light limited due to shading by other algae, aquatic vascular plants and even terrestrial vegetation. Currents, wave action, land runoff or animal activity also can deprive microalgae of light by increasing the turbidity of the water and by covering the algae with sediment. Less than 1 mm of muddy sediment will reduce the light intensity to below the compensation intensity. The littoral zone is much more subject to extreme and short-term fluctuations in environmental parameters than open-water habitats. Benthic algae must be able to withstand and to recover rapidly from temperature, desiccation and/or salinity shocks. In ephemeral freshwater ponds they must survive dry spells. In the densely populated littoral community, competition for space is likely to become severe and

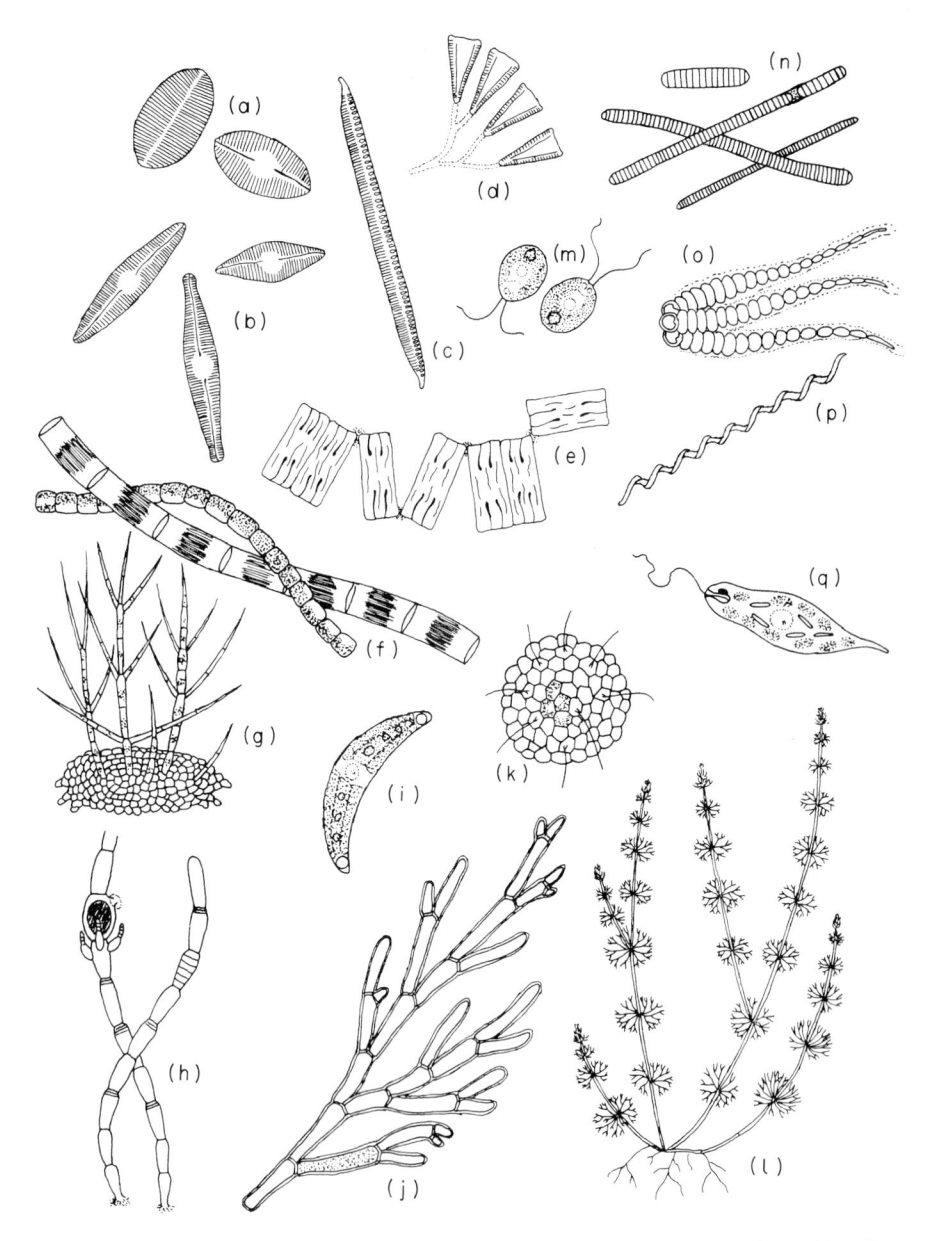

Fig. 5.1 Examples of benthic algae. Diatoms: (a) *Cocconeis*; (b) *Navicula*; (c) *Nitzschia*; (d) *Gomphonema*; and (e) *Tabellaria*. Green algae: (f) *Ulothrix*; (g) *Stigeoclonium*; (h) *Oedogonium*; (i) *Closterium*; (j) *Cladophora*; (k) *Coleochaete*; (l) *Nitella*; and (m) *Chlamydomonas*. Blue-green algae: (n) *Oscillatoria*; (o) *Calothrix*; and (p) *Spirulina*. Euglenoids: (q) *Euglena*. (Drawings by Andrew J. Lampkin III.)

other biotic interactions among algae, bacteria, fungi, grazing organisms and vascular plants become much more complex than they are in open water.

To understand the physiological ecology of benthic algae, we need to identify the algae present, quantify their standing crop and measure their rate functions. In addition, we must know something about their environment and be able to experimentally manipulate it in defined and reproducible ways in our research effort. These tasks are incredibly more difficult and complicated in the littoral zone than they are in open water. The problem is especially acute in the case of benthic microalgae which live in an intimate and complex relationship with their substrate. They exist in a boundary layer between a relatively homogeneous body of water and some other environment. In the case of a hard, impenetrable surface such as a rock, that other environment is modified by the adsorptive properties of that particular surface and by the slimy bacterial film that rapidly develops on it. A vascular plant would be expected to be a much more contrasting and fluctuating second environment which could actively modify the attached alga's chemical environment in a number of ways. Microalgae living at the surface of muddy sediments are in perhaps the most extreme gradient of all. In a matter of millimetres the environment changes from oxidizing to reducing with profound effects on the abiotic and biotic components of the system.

The first difficulties in studying these algae are encountered in simply attempting to obtain representative and replicate samples. Even if a given unit area can be accurately identified and sampled in this rugged (on a microscale) terrain, the extreme patchiness in algal distribution makes it difficult to obtain reliable data and to interpret them. The settling of planktonic species further complicates the picture. Standing-crop estimates based on chlorophyll determinations are subject to contamination by relatively high levels of chlorophyll-degradation products in sediments and by chlorophyll in living plant substrates.

Measurements of rate functions such as algal photosynthesis are complicated by the presence of other autotrophs and by the activity of the large standing crop of heterotrophs whose respiration could mask low net photosynthetic rates when measured by the O_2 method. Radiocarbon productivity measurements are complicated by the difficulty of counting this low energy beta emitter in the presence of particulate matter. It is difficult to enclose representative samples of the system for productivity measurements without modifying the environment and therefore changing photosynthetic rates. Since it is virtually impossible to adequately define or reproduce the microhabitat of these cells, results of experimental manipulations of the environment are difficult to interpret. For example, how do the ion exchange characteristics of the substrate affect dissolved nutrient concentrations in the microzone surrounding the cell? How does the redox potential of the system affect nutrient supply? Is the nutrient level next to the cell increased in direct proportion to experimental increases in the nutrient concentrations in the nearby water? Such uncertainties make it difficult to extrapolate from laboratory studies of cultures to the alga's natural environment. It is easier to define the environment of the larger filamentous algae which extend into the water from an attachment point, although at some crucial point in their life cycle a microscopic spore or cell had to settle down and become established in this same complex, boundary layer at the substrate surface. The preceding discussion is presented more as a cautionary note against simplistic approaches to this system than as an argument that any meaningful research is hopeless.

Compared with planktonic algae, very little is known about the autecology of benthic algae. The reasons are related to the difficulties mentioned in the preceding paragraphs rather than to any lack of importance of these algae to their systems. Littoral algae in lakes, for example, can contribute over half of the total annual production in some lakes, depending on the lake morphometry and substrate characteristics. Most ecological studies on benthic algae are floristic and correlative in nature with very few examples of experimental tests of suspected controlling influences. Where information is lacking, we can begin with the premise that benthic algae respond to environmental parameters in much the same way as planktonic species. We must now devote more attention to testing hypotheses developed from the descriptive studies to discover any unique physiological, structural and reproductive adaptations which may have evolved in response to the benthic environment.

PERIPHYTON

Periphyton is used here as a general term to refer to algae living on some type of solid substrate such as rock, wood, other plants or various manmade objects. It is essentially synonymous with the German word, Aufwuchs. We will consider the two most common categories of periphyton: algae living on rocks (epilithic) and those living on vascular plants or on other algae (epiphytic). The floras of these two habitats have many species in common and are relatively distinct from the epipelic assemblage considered later.

Epilithic algae

The epilithic community usually develops in areas where water movements are sufficiently strong to prevent finer sediments from accumulating; streams and rivers and rocky shores are the best-studied habitats. The algae found here have various, well-developed attachment mechanisms including: a series of prostrate filaments which support erect, branching filaments in the green alga *Stigeoclonium*; rhizoidal extensions of the basal cell in *Cladophora*; a single, relatively undifferentiated, holdfast cell in the unbranched, green filaments *Ulothrix* and *Oedogonium*; and the secretion of a cushion, stalk or tube of gelatinous material by many species of pennate diatoms.

In streams and rivers, light intensity and current flow are usually considered the dominant environmental factors influencing the algae. In many streams a definite increase in the periphyton is observed in the spring and is ascribed to increasing light and possibly temperature increases as well. The summer decline in algal abundance is not as well correlated with nutrient deficiency as it is in lakes but may be due to high temperature or grazing pressure or to decreasing light intensities as the tree canopy fills in above small streams. Recently, however, nutrient enrichment experiments have demonstrated that nutrient (especially phosphorus) limitation may be important in certain streams (Elwood *et al.* 1981). Seasonal succession in species abundance is just as apparent in periphyton as it is in phytoplankton. In some streams the scouring action of periodic increases in river flow caused by heavy rain may drastically reduce the standing stock and obscure trends due to other factors. In rivers and streams the wash-out from the benthic algal assemblage generally accounts for most of the suspended algae in the water, although slow-moving rivers develop a true phytoplankton dominated by small centric diatoms and small flagellates.

Beyond the physical effects of scouring and wash-out, there are also physiological effects of current on epilithic algae. Certain algae, freshwater reds for example (Sheath & Hymes 1980), are only found growing in flowing water. In experimental laboratory streams, filamentous algae (*Stigeoclonium*, *Oedogonium*, *Tribonema*) dominated the periphytic community when the flow was held at 9 cm s^{-1} whereas a dense, felt-like growth of diatoms dominated streams with a flow rate of 38 cm s^{-1} (McIntire 1966). A number of studies (e.g. Lock & John 1979) have shown that rates of growth, nutrient uptake, respiration and photosynthesis are higher in moderate currents than in slow currents or in still water. Currents increase metabolic rates by creating steeper diffusion gradients next to the cell. In an interesting comparison of related species of filamentous green algae characteristic of stream and lake habitats, Whitford and Schumacher (1964) found that the stream species had the more pronounced response to current. The uptake rate of phosphate by *Oedogonium kurzii*, a stream species, was stimulated 16 fold by a current of 20 cm s^{-1}, whereas the same current stimulated uptake by an unidentified lake species of *Oedogonium* only 7.4 fold. The authors concluded that stream species tend to have higher metabolic rates than lake species and therefore require the more rapid exchange of metabolites provided by currents.

The nature of the substrate seems to be less important than other environmental parameters in determining the spatial distribution of algae, although some differences have been noted in the communities developing on different types of rocks. Light intensity is known to affect the distribution of species which prefer shady situations (e.g. the red alga *Batrachospermum*) or sunny spots (e.g. *Cladophora*). Higher temperatures (above 30–35 °C) may result in an increase in the abundance of blue-greens at the expense of diatoms. Along the wave-washed shores of lakes, one finds a definite zonation of benthic microalgae; the forms which can withstand periodic desiccation, especially blue-green algae, are found at higher levels than forms which are more sensitive to exposure.

Artificial substrates such as glass microscope slides are commonly employed in the study of epilithic microalgae because they are colonized within two to ten weeks by an assemblage of algae which is reasonably representative of natural epilithic communities. The community structure of periphyton developing on artificial substrates is more representative of the natural assemblage in areas where the natural substrate is periodically cleaned by natural forces (scouring, desiccation). In areas subject to less perturbation (the sublittoral zone of lakes) the natural periphyton community does not readily develop on a fresh substrate (Loeb 1981). Since artificial substrates provide a more uniform and reproducible habitat with less patchiness in algal distribution than natural substrates, sampling problems are not as formidable and the lower variance in the data makes it easier to interpret the results of experimental manipulations. Styrofoam or polyethylene sheeting are also very good substrates and have the advantage that subsamples can be obtained easily with a cork borer. The large number of studies utilizing artificial substrates almost always has emphasized community-level features such as species diversity or primary productivity rather than the autecology of the individual algal species.

It seems clear that major advances in our understanding of epilithic algae are possible using a combination of presently available laboratory and field techniques. As a first approach a comprehensive comparison of the physiology of benthic and planktonic algae in suspension culture would be useful. For example, do nutrient uptake kinetics differ significantly between the two groups? In more realistic experiments, cultures of pennate diatoms or other epilithic algae could be grown attached to an easily sampled substrate (agar, styrofoam, polyethylene sheeting, dialysis membranes) in some sort of flow-through system to study the effects of current and other variables on cell physiology and growth rate. Why not transplant laboratory-colonized substrates to the field under various conditions to help understand seasonality and competition in epilithic algae? Are certain species restricted to rapidly flowing water by competition, grazing pressure, temperature, gas exchange, nutrient requirements or the build up of toxic metabolites? Why are certain species able to withstand periodic desiccation and others are not? A very important aspect of the epilithic habitat is the organic layer that rapidly forms on submerged surfaces. It is apparently derived from bacterial secretions and adsorbed, dissolved, organic matter. Bacterized systems can be used to study its role, if any, in the settling, attachment and growth of periphyton. Much of this research can take advantage of the fact that a periphytic algal assemblage is localized, easily moved from spot to spot and recoverable to a limited extent. In certain respects it can be treated as if it were a dialysis culture.

Epiphytic algae

It is perhaps surprising that many of the observations and concepts discussed with respect to epilithic algae also apply to the epiphytic algae growing on vascular plants and other algae. One might expect that the differences between a living and a non-living substrate would have profound effects on the respective periphytic communities. In fact, epiphytic algae generally show little substrate specificity; many epiphytic species are encountered in natural epilithic communities and on artificial substrates. Eminson and Moss (1980) have suggested that host specificity is more likely to occur in oligotrophic than in eutrophic habitats since chemical interactions between host and epiphyte leading to specificity would be masked under nutrient-rich conditions.

In spite of the seeming relative indifference of epiphytic algae to their substrate, the epiphytic habitat has several distinctive attributes. The surface itself has a definite life span. New leaves are colonized as they develop during the growing season resulting in a summer or autumn peak in epiphyte biomass and productivity. Artificial substrates resemble leaves to the extent that they are also fresh, colonizable surfaces although they obviously do not provide the same environment as a living plant. The canopy of aquatic macrophytes often creates light-limiting conditions for epiphytic algae.

Much of the physiological research on epiphytic algae has focused on the

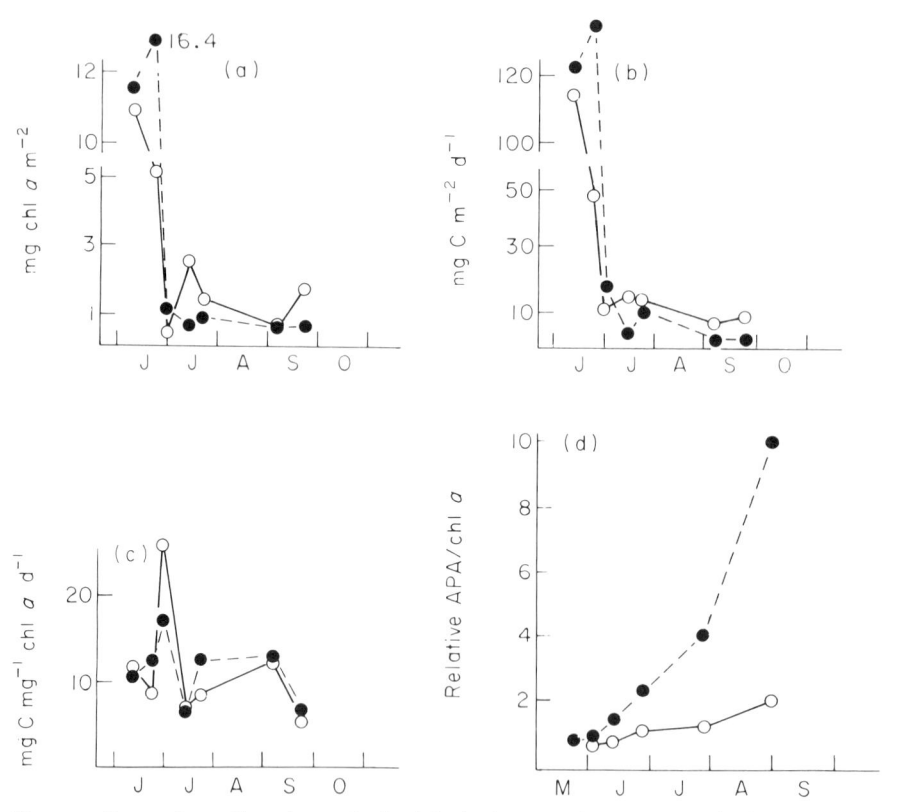

Fig. 5.2 Comparison of loosely attached epiphytic algae growing on a natural substrate (open circle) and on an artificial substrate (solid circle). (a) Algal biomass per unit area of host plant. (b) Productivity of epiphytes per unit area of host plant. (c) Assimilation number (see p. 31). (d) Alkaline phosphatase activity (APA) relative to the maximum observed value. (After Cattaneo & Kalff 1979.)

question of nutrient transfer from rooted, aquatic, vascular plants to their epiphytes. A few studies have demonstrated a transfer of organic carbon, nitrogen or phosphorus from macrophyte to the epiphytic community. There is evidence that some rooted aquatic plants act as pumps, transferring phosphorus and other nutrients from the sediments to epiphytes and the water column (Penhale & Thayer 1980). The amount of nutrient released, however, is only a few per cent of that utilized by the plant. This minor role of nutrient transfer was demonstrated in a Canadian lake by comparing epiphytic growth on *Potamogeton* plants and morphologically similar plastic plants (Cattaneo & Kalff 1979). The comparison was limited to those epiphytes that were loosely attached (removed by vigorous shaking). As shown in Fig. 5.2, there was no difference in biomass (as chlorophyll *a*), productivity or assimilation number, suggesting this living plant is a neutral substrate for the epiphytes. Some small transfer of phosphorus is suggested, however, by the lower levels of alkaline phosphatase activity (see p. 42) in the epiphytic community on the living plants.

A number of algae have established more intimate interactions with their hosts as obligate epiphytes or as parasites. Red algae living on other red algae are the best-studied examples. *Janczewskia gardneri* is a small, lightly pigmented epiphyte on *Laurencia spectabilis* to which it is attached by filaments which penetrate the host. Although these characteristics suggest a parasitic relationship, experiments with ^{14}C show that both algae are capable of photosynthesis and that there is no detectable translocation of organic carbon between partners (Court 1980). *J. gardneri* appears to be an obligate epiphyte. In another well-studied example, *Harveyella mirabilis* occurs as a small, whitish parasite on several different red algae including *Odonthalia floccosa*. Radiotracer studies (Fig. 5.3) show that photo-

Fig. 5.3 Changes in levels of ^{14}C in prelabeled infected and uninfected *Odonthalia* and in *Harveyella* with and without the subtending host tissue. The plants were illuminated during this experiment. (After Goff 1979.)

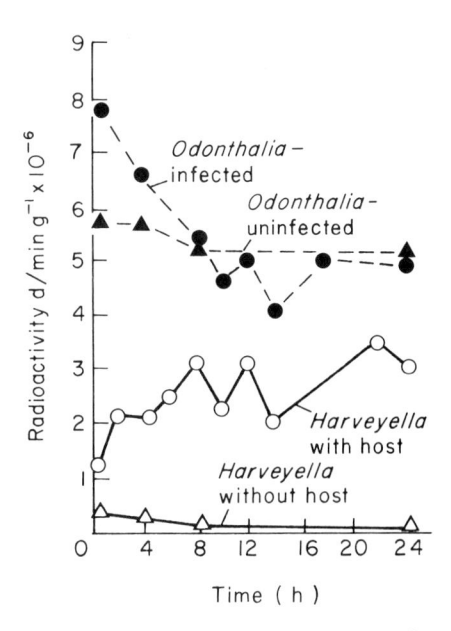

synthetically fixed carbon is translocated from *O. floccasa* to *H. mirabilis* (Goff 1979). The transfer takes place in a zone of interdigitation of host medullary cells and parasite rhizoidal cells.

EPIPELIC ALGAE

The epipelic flora consists of the algae living in or on the surface of soft, muddy sediments in intertidal mud-flats or salt-marshes, the littoral zone of lakes and in areas of slow-moving water in streams and rivers. This benthic algal assemblage is also dominated by pennate diatoms and blue-green algae although euglenoids and motile species of other algal classes are also encountered. Motility is a definite selective advantage in this habitat due to the danger of being buried by the muddy sediments. Although not actually part of the epipelic flora, a transitory assemblage of algae, especially desmids, *Spirogyra* and other members of the Zygnematales, will often develop near the sediment surface in areas protected from swift currents. Epipsammic algae are found in sandy sediments and will be included here although some authorities limit the term to those small cells (diatoms and blue-greens) which are firmly attached to sand grains.

The ecology of epipelic algae is especially difficult to study due to the intimate association of the microalgae with the soft, easily disturbed sediments. Periphyton is relatively easy to scrape from its substrate with a minimum of contamination by extraneous material. Any scraped sample of epipelic algae, however, contains a great deal of sedimentary material which obscures microscopic observations as well as practically any other measurement one would care to make. Another vexing, albeit fascinating and even useful, aspect of epipelic algae is their well-documented, vertical-migration behaviour within the upper 0.2–4.0 mm of sediment (Harper 1977). The cells (pennate diatoms, euglenoids and blue-green algae) generally migrate to the sediment surface at dawn and return at sunset, although in freshwater ponds they may retreat from bright, midday light intensities. The diel migratory behaviour is thought to be regulated by circadian rhythms of phototaxis and/or geotaxis which may be retained in the laboratory for several days. Thus, in the morning the cells exhibit positive phototaxis and negative geotaxis in their migration to the surface while the reverse responses prevail in the evening hours. In intertidal sediments, the migratory response is modified so that the cells appear on the surface during daytime low tides. In most species examined the tidal aspect of the rhythm appears to be maintained by some aspect of the tidal regime (perhaps changes in light intensity, water content of the sediment or pressure) since it disappears rapidly when sediment is brought into the laboratory. In the diatom *Hantzschia virgata*, however, the tidal rhythm continues in the laboratory under constant light, keeping pace with the time of low tide which advances approximately 50 min each day (Fig. 5.4). Even more amazing was the finding that the cells rephased their cycle (in the laboratory) to pick up the morning low tide once the afternoon low tide advanced into the evening hours (Palmer & Round

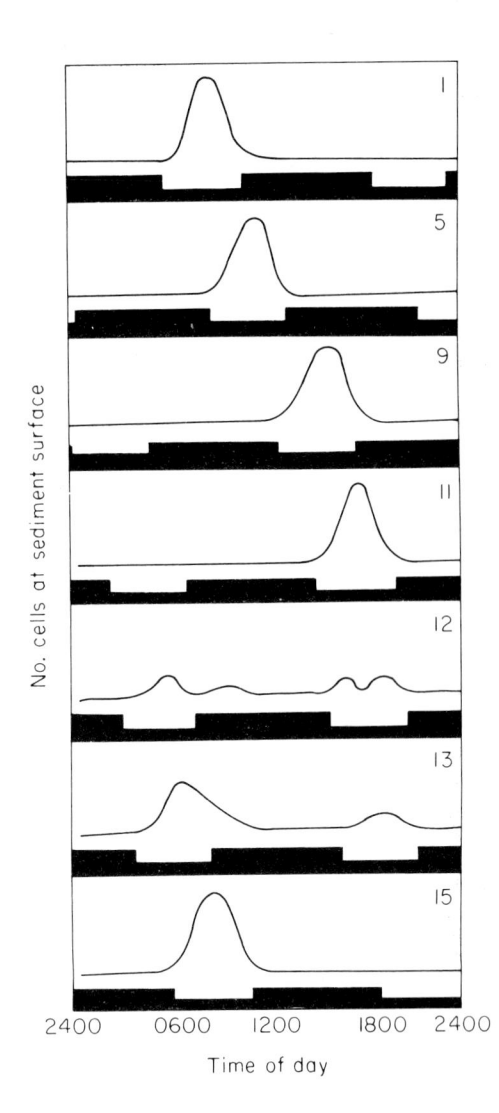

Fig. 5.4 Vertical migration rhythms in *Hantzschia virgata* after the transfer to continuous-light conditions in the laboratory. The bar graph at the bottom of each panel indicates the time of the high and low tides in the field. The number at the top right of each panel indicates the number of days since the samples were brought into the laboratory. The natural photoperiod at the time of collection was approximately 0530–2000 h. (After Palmer & Round 1967.)

1967). The ecological significance of the downward migration during periods of darkness is not clear, but could be related to predator avoidance or nutrient assimilation.

Quantitative data on vertical migration rhythms are obtained by placing glass cover slips or two layers of lens paper or bolting cloth on the sediment surface for several hours. As cells migrate to and from the surface the number of cells attached to these easily removable materials varies accordingly. One can also take advantage of this system to obtain clean suspensions of epipelic algae for various laboratory experiments by simply recovering the top layer of lens paper or cloth and rinsing off the cells. There are problems with this technique in that the natural environment of the cells is completely altered and that recovery of the epipelic algal assemblage is not complete and may not be representative of the entire flora. This approach is an improvement over using cultured material, however, because the physiol-

83

ogy and chemical composition of field-grown cells can be studied without the complications caused by the sediments.

The study of primary production in epipelic algae has been approached in many different ways. A few investigators have used dilute suspensions of sediment samples to measure 'potential' photosynthesis, whereas most have tried to maintain the natural integrity of the alga–substrate relationship to obtain a more realistic estimate of actual productivity. These latter techniques are usually modifications of ^{14}C or O_2 methods and use core samples of sediments, incubated in the field or laboratory. Most methods employ submerged conditions even though intertidal epipelic algae are at the surface and presumably most productive under exposed (low tide) conditions. The interesting results from productivity studies on intertidal seaweeds (see Chapter 6) should encourage a comparison of photosynthetic rates of epipelic algae under exposed and submerged conditions.

The distribution of epipelic and epipsammic algae appears to be more dependent on the physical and chemical properties of the substrate than is the case for the periphytic flora, although water chemistry and temperature continue to be important variables. Light intensity is often limiting due to the shade from vascular plants or turbid estuarine water. Since the depth of the 1% light level may vary from about 0.2 mm in mud to 3.0 mm in sand, sediment movements which bury the algae can be limiting through their effect on light intensity. The demonstrated ability of several epipelic diatoms to secrete mucilage which consolidates and stabilizes sediments could be an adaptive mechanism to help reduce the resuspension of sediments (Holland *et al.* 1974).

The effect of high light intensities is also of interest since the migration behaviour of intertidal species exposes them to full sunlight which would severely inhibit photosynthesis in planktonic species. Available evidence from both *in situ* studies and suspensions of cells collected by the lens-paper technique shows that the P versus I curve of these diatoms has a broad optimum, saturating at fairly low light intensities with only slight inhibition under full sunlight. Vertical migration rhythms obviously complicate productivity measurements in undisturbed sediments since the time of day or any experimental manipulation which affects migration also will affect the light intensity experienced by the cells and therefore may affect photosynthetic rates. Different workers have ascribed diel rhythms in photosynthetic rates of *in situ* epipelic algae to either endogenous rhythms in photosynthetic capacity or to the light intensity effects of migration.

Inorganic nutrients are not considered to be limiting factors for the growth of algae in most epipelic habitats due to the relatively high nutrient concentrations in sediments. It is possible, however, that localized nutrient deficiency could occur during blooms when other parameters (light, temperature) are optimum. Nitrogen limitation of epipelic algae has been demonstrated by nutrient-enrichment experiments in salt-marshes along the eastern coast of the USA (Darley *et al.* 1981). It is likely that marsh grasses, which are also nitrogen limited, are competing for the available nitrogen. It seems worthwhile to investigate the possibility that O_2 tensions are low enough in the surface sediments to permit N_2 fixation in non-heterocystous blue-green algae. There are few studies of nutrient-uptake kinetics in epipelic algae; one study (Admiraal 1977a) reported a K_s of c. 0.1 μmol l^{-1} for phosphate-limited growth in a benthic diatom. That this value is similar to those reported for planktonic algae suggests that benthic algae may retain their high affinity for nutrients in a generally nutrient-rich environment. More species obviously need to be investigated.

The intertidal zone is subject to extreme and rapid environmental changes, particularly with respect to temperature, salinity and desiccation. The intertidal seaweeds have been much more extensively studied in this regard than the microalgae. Cultures as well as natural populations of epipelic diatoms are much more tolerant of salinity changes than are planktonic (species (Admiraal 1977b). In the salinity range of 4–60% photosynthetic rates remained within 30% of the initial values for up to six hours. Algae collected from sites with mean salinities of 5–30% showed only slight differences in salinity response demonstrating that salinity may not be as important in controlling distribution as correlative studies in the field have suggested. To date no significant differences in temperature response between intertidal microalgae and phytoplankton have been reported.

Frequent light-limiting conditions and high concentrations of dissolved organic carbon in epipelic habitats suggest that facultative heterotrophy would be a selective advantage for epipelic algae. Indeed, many epipelic pennate diatoms are capable of heterotrophic growth in culture. The challenge is to obtain evidence that heterotrophy is actually of significance to natural populations of epipelic algae (see p. 50). We know very little about the concentrations of usable substrates in sediments, particularly in the surface 1–2 mm where these algae live, but concentrations are possibly in the neighbourhood of 1 μmol l^{-1}. One recent study (Darley et al. 1979) measured the assimilation of ^{14}C-labeled substrates at 1 μmol l^{-1} concentrations by field-grown diatoms collected from a salt-marsh mud-flat by the lens-paper technique. At the specific carbon-incorporation rates observed, they calculate that the cells could obtain less than 1% of their carbon heterotrophically if they had generation times (a doubling of cell carbon) of 48 hours. It is possible that this small amount of additional carbon is enough to provide a competitive advantage in light-limiting conditions or to enhance cell survival during prolonged burial in the sediment.

CHARALES

The charophytes (e.g. *Chara*, *Nitella*) may form substantial populations in the littoral zone of lakes and in some streams and rivers, particularly in hard water of alkaline pH (Hutchinson 1975). Their sensitivity to high phosphate levels apparently limits their occurrence in eutrophic lakes. The plants are anchored in the silty or muddy (less often sandy) substratum by colourless rhizoids and for this reason the charophytes are in competition with vascular aquatic plants for space and other resources. In general, the charophytes appear to be the losers in this competition. They often appear in newly formed ponds or lakes only to be replaced by angiosperms as the system matures. The low light intensities resulting from the canopy produced by angiosperms often restricts charophytes to deeper water where angiosperms cannot grow. As a group, charophytes appear to be more shade tolerant than aquatic angiosperms, possibly because they have less non-photosynthetic tissue to maintain. The absence of the pressure-sensitive, intercellular, gas system found in aquatic vascular plants no doubt also contributes to the relative success of charophytes in deeper waters. Mechanical stress of wave action in shallow exposed areas can be a limiting factor to both charophytes and vascular plants.

The large internodal cells of the charophytes have been used in many experimental studies on the physiology of ion transport phenomena, but relatively little effort has been devoted to the experimental physiological ecology of the group. It is known that phosphate taken up by the rhizoids is translocated to the rest of the plant, presumably by cytoplasmic streaming within internodal cells and diffusion through plasmodesmata between cells. Although other parts of the plant are equally effective at phosphate uptake, it is possible that the nutrient-rich sediments could be a major source of nutrients. Plants suspended above the sediment, with or without attached rhizoids, grow more slowly than plants which are rooted in the mud. There must also be a net movement of O_2 and reduced carbon to the rhizoidal system to support growth in the generally anoxic sediments.

REFERENCES

ADMIRAAL W. (1977a) Influence of various concentrations of orthophosphate on the division rate of an estuarine benthic diatom, *Navicula arenaria*, in culture. *Mar. Biol.*, **42**, 1–8.

ADMIRAAL W. (1977b) Salinity tolerance of benthic estuarine diatoms as tested with a rapid polarographic measurement of photosynthesis. *Mar. Biol.*, **39**, 11–18.

CATTANEO A. & KALFF J. (1979) Primary production of algae growing on natural and artificial aquatic plants: A study of interactions between epiphytes and their substrate. *Limnol. Oceanogr.*, **24**, 1031–7.

COURT G.J. (1980) Photosynthesis and translocation studies of *Laurencia spectabilis* and its symbiont *Janczewskia gardneri* (Rhodophyceae). *J. Phycol.*, **16**, 270–9.

DARLEY W.M., MONTAGUE C.L., PLUMLEY F.G., SAGE W.W. & PSALIDAS A.T. (1981) Factors limiting edaphic algal biomass and productivity in a Georgia salt-marsh. *J. Phycol.*, **17**, 122–8.

DARLEY W.M., OHLMAN C.T. & WIMPEE B.B. (1979) Utilization of dissolved organic carbon by natural populations of epibenthic salt-marsh diatoms. *J. Phycol.*, **15**, 1–5.

ELWOOD J.W., NEWBOLD J.D., TRIMBLE A.F. & STARK R.W. (1981) The limiting role of phosphorus in a woodland stream ecosystem: Effects of P enrichment on leaf decomposition and primary producers. *Ecology*, **62**, 146–58.

EMINSON D. & MOSS B. (1980) The composition and ecology of periphyton communities in freshwaters. 1. The influence of host type and external environment on community composition. *Br. phycol. J.*, **15**, 429–46.

GOFF L.J. (1979) The biology of *Harveyella mirabilis* (Cryptonemiales, Rhodophyceae). VI. Translocation of photoassimilated ^{14}C. *J. Phycol.*, **15**, 82–7.

HARPER M.A. (1977) Movements. In Werner D. (ed.), *The Biology of Diatoms*, pp. 224–49. Blackwell Scientific Publications, Oxford.

HOLLAND A.F., ZINGMARK R.G. & DEAN J.M. (1974) Quantitative evidence concerning the stabilization of sediments by marine benthic diatoms. *Mar. Biol.*, **27**, 191–6.

HUTCHINSON G.E. (1975) *A Treatise on Limnology. III. Limnological Botany.* John Wiley & Sons, New York.

LOCK M.A. & JOHN P.H. (1979) The effect of flow patterns on uptake of phosphorus by river periphyton. *Limnol. Oceanogr.*, **24**, 376–83.

LOEB S.L. (1981) An *in situ* method for measuring the primary productivity and standing crop of the epilithic periphyton community in lentic systems. *Limnol. Oceanogr.*, **26**, 394–9.

MCINTIRE C.D. (1966) Some effects of current velocity on periphyton communities in laboratory streams. *Hydrobiologia*, **27**, 559–70.

MCINTIRE C.D. & MOORE W.W. (1977) Marine littoral diatoms: Ecological considerations. In Werner D. (ed.), *The Biology of Diatoms*, pp. 333–71. Blackwell Scientific Publications, Oxford.

PALMER J.D. & ROUND F.E. (1967) Persistent, vertical-migration rhythms in benthic microflora. VI. The tidal and diurnal nature of the rhythm in the diatom *Hantzschia virgata*. *Biol. Bull.*, 132, 45–55.

PATRICK R. (1977) Ecology of freshwater diatoms and diatom communities. In Werner D. (ed.), *The Biology of Diatoms*, pp. 284–332. Blackwell Scientific Publications, Oxford.

PENHALE P.A. & THAYER G.W. (1980) Uptake and transfer of carbon and phosphorus by eelgrass (*Zostera marina* L.) and its epiphytes. *J. exp. mar. Biol. Ecol.*, **42**, 113–23.

SHEATH R.G. & HYMES B.J. (1980) A preliminary investigation of the freshwater red algae in streams of southern Ontario, Canada. *Can. J. Bot.*, **58**, 1295–318.

WETZEL R.G. (1975) *Limnology.* W.B. Saunders Co., Philadelphia.

WHITFORD L.A. & SCHUMACHER G.J. (1964) Effect of a current on respiration and mineral uptake in *Spirogyra* and *Oedogonium*. *Ecology*, **45**, 168–70.

WHITTON B.A. (ed.) (1975) *River Ecology.* University of California Press, Berkeley, California.

6 Seaweeds

In this chapter we will consider the physiological ecology of macroscopic green, brown and red algae (Fig. 6.1) that grow in the intertidal and subtidal zones of the seashore (see Chapman 1979; Dawson 1966). The plants are almost always attached, usually to a stable substrate and are therefore most abundant on rocky shores or on man-made substrates such as breakwaters, seawalls, pilings, etc. Attachment structures, or holdfasts, vary from rhizoidal extensions of basal cells, to prostrate, filamentous discs, to the specialized haptera of the kelps (Laminariales of the brown algae). Seaweeds range in size from small and easily overlooked crustose forms and uniseriate filaments to the giant kelps. Some seaweeds are perennials, even though they may be reduced to a basal holdfast and stipe during the off-season, while others appear to be annuals but actually exist in a less conspicuous form, often an alternate phase of the life cycle, during the winter months. In an extreme case of longevity, an intertidal, crustose red alga was reported to have a 4% increase in surface area per year; calculated ages of individual plants approached 100 years (Paine et al. 1979). Few seaweeds produce any kind of resistant spore or resting stage. In addition to the physiological stress of the intertidal zone, the algae must be well anchored and tough yet flexible and resilient to withstand the mechanical stress of wave action. Some larger brown algae produce air bladders that help suspend the photosynthetic tissues nearer to the water's surface.

Early research on the physiological ecology of seaweeds emphasized the ability of intertidal species to survive exposure to extremes of temperature, salinity and desiccation. More recently, rate functions, including comparisons during emergence and submergence, have been receiving attention. The availability of scuba diving gear is allowing researchers to study subtidal species *in situ*. The size of the plants is a mixed blessing for research. On the one hand, biomass estimates and productivity measurements based on them are relatively straightforward due to the ease of handling and harvesting macroscopic plants. The longevity of most plants permits repeated observations on the same material for studies of seasonal adaptation. Furthermore, plants can be moved for experimental purposes. On the other hand, culturing large seaweeds can be an expensive operation requiring large tanks and a great deal of plumbing to provide running sea-water. Of necessity, therefore, many studies of seasonal physiology utilize field-grown

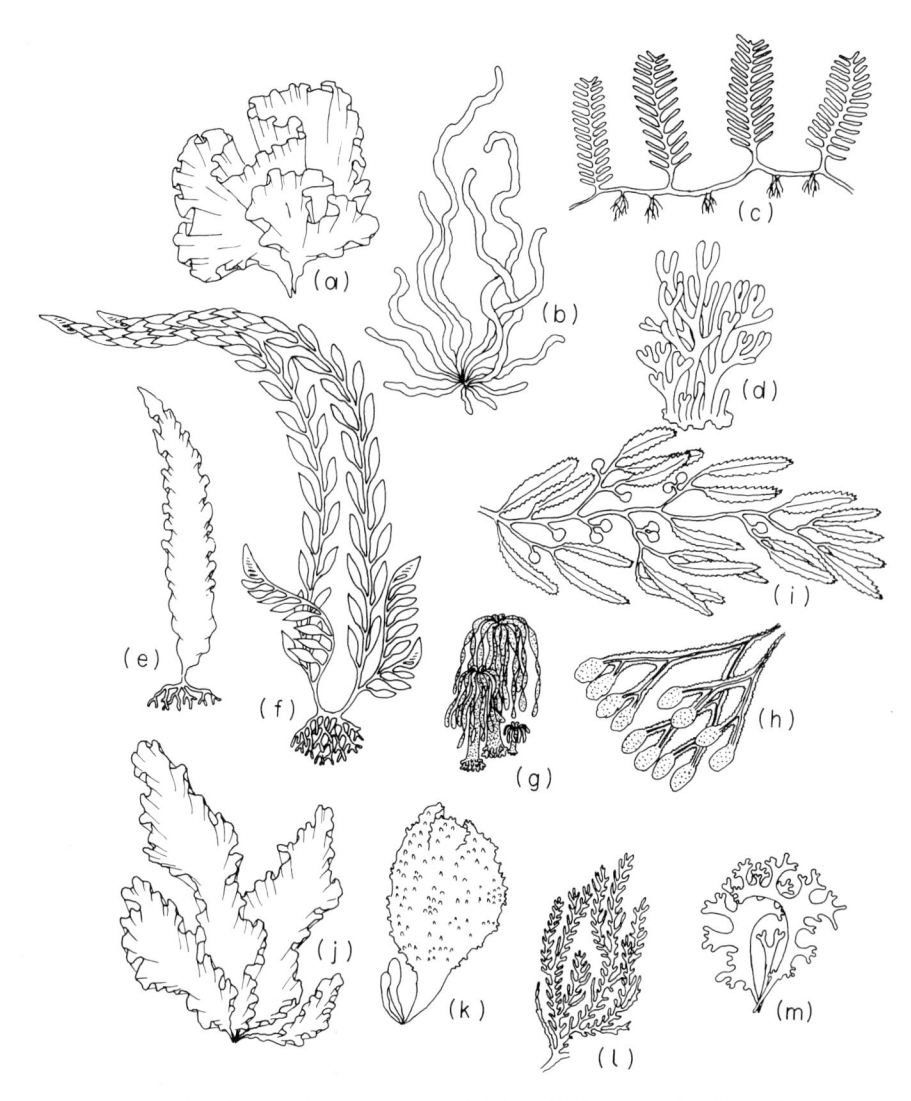

Fig. 6.1 Examples of seaweeds. Green algae: (a) *Ulva*; (b) *Enteromorpha*; (c) *Caulerpa*; and (d) *Codium*. Brown algae: (e) *Laminaria*; (f) *Macrocystis*; (g) *Postelsia*; (h) *Fucus*; and (i) *Sargassum*. Red algae: (j) *Porphyra*; (k) *Gigartina*; (l) *Gelidium*; and (m) *Chondrus*. (Drawings by Andrew J. Lampkin III.)

material in which it is difficult to identify physiological adaptations to individual environmental parameters. An additional problem is that the rate measurements in many seasonal studies have been limited to ambient conditions. While this approach is very useful for data on *in situ* performance, it provides little insight to possible mechanisms of seasonal adaptation. Physiological studies should also take into account plant to plant variations as well as the differences that exist in different parts of the plant. Epiphytic growth on seaweeds may interfere with some studies, but unialgal culture of the larger forms is difficult.

Seaweed productivity on a unit area basis, particularly in the subtidal kelp forests, may equal vascular plant production in some of the most productive terrestrial systems (Mann 1973). One estimate places worldwide seaweed production as high as 10% of total oceanic production. Numerous techniques have been used to measure seaweed productivity. Measurements based on biomass are more convenient but may be affected by herbivore grazing and by the tendency of older plant parts to erode. In kelps which have intercalary meristems at the junction of the stipe and blade, the rate of growth can be estimated by the rate at which a marker (a hole) moves along the blade (Fig. 6.2). Most productivity estimates are obtained by following dissolved oxygen in the standard light–dark bottle method. It is important to use small amounts of seaweed material to avoid creating stress conditions in the bottles during incubation. Thus, whole plants can be used only for the smallest species whereas branches, growing tips or excised discs or strips are used for larger species. Using only part of the plant can lead to errors because younger, growing regions of larger species usually exhibit higher carbon fixation per unit biomass. Cutting the thallus can also lead to large increases in respiration rates. 'Uprooting' attached seaweeds apparently has no effect on productivity as long as tissue damage is kept to a minimum. Even brief exposure of subtidal algae to air and bright sunlight, however, can seriously inhibit productivity for several weeks or more. Hatcher (1977) describes a diver-operated apparatus for measuring productivity of whole plants *in situ*. The plants can be anchored in place following the measurement and used again at a later date. Radiocarbon is used less frequently in seaweed productivity measurements because of the interference of the tissue with subsequent counting procedures. Photosynthesis in air (low tide conditions) can be measured with the gas exchange systems used for higher

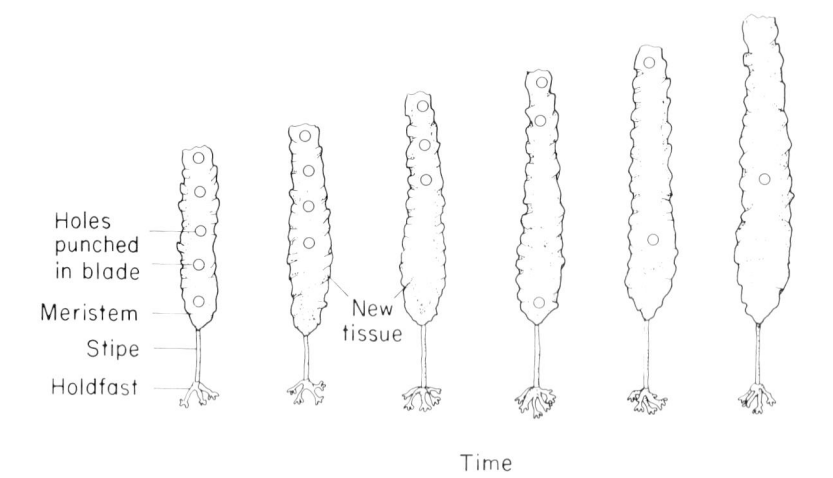

Fig. 6.2 In kelps which have a localized meristem, the growth rate may be obtained by measuring the tissue added between the meristem and the most recently punched hole. The growth rate is faster than the rate of increase of the blade length because the blade is eroding at the tip. (After Mann 1973.)

plants. Changes in the CO_2 concentration are recorded with an infra-red gas analyser as the air passes through a transparent, environmentally controlled chamber containing the algae.

ENVIRONMENTAL FACTORS AFFECTING GROWTH

Perennial as well as annual seaweeds show definite seasonal cycles in growth and reproduction which have been correlated with seasonal fluctuations in light intensity, temperature and less often nutrient concentration. Growth rates usually begin to increase in late winter or spring and decline in the summer or autumn. Photosynthetic rates measured under ambient environmental conditions show corresponding changes during the year. Available data indicate that changing photosynthetic rates in perennial species are imposed by the existing environmental conditions rather than by intrinsic changes in the actual photosynthetic capacity of the alga itself. Some seasonal adaptation is apparent, however (see p. 97).

One of the most complete studies on the seasonal growth of a seaweed is that on the kelp, *Laminaria longicruris*, of Nova Scotia, Canada. Although growth continues throughout the year, there is a sharp increase in growth rate in the winter months when low temperatures and light intensities might be expected to severely restrict growth. An explanation for this anomalous behaviour has been provided in a series of papers (see Chapman & Craigie 1978); some of the pertinent data are presented in Fig. 6.3. The summer decline in growth rate is due to nitrogen depletion as shown by large-scale, nutrient-enrichment experiments. (Three kg of $NaNO_3$ was placed in porous clay pots from which it diffused within four to seven days in the field.) In the annual growth cycle the NO_3^- and organic nitrogen contents of the plants increase following the normal winter increase in NO_3^- in the water and the plants take advantage of their improved nitrogen status to initiate growth. Growth during December and January must be partially at the expense of carbon reserves (laminaran and mannitol) since net photosynthetic carbon fixation is not sufficient (limited by light and temperature) to account for the carbon found in new tissue. After January net photosynthesis is sufficient to support growth. The plants are able to utilize internal stores of NO_3^- to maintain high growth rates for several weeks beyond the point of NO_3^- depletion in the water. Following the depletion of stored NO_3^-, the organic nitrogen in the plant declines as growth rate drops in the summer. Laminaran and mannitol reserves begin to accumulate once assimilated carbon is diverted from new growth by nitrogen depletion. Thus the ability to store carbon and nitrogen in the large thallus for later use and the ability to photosynthesize at low temperatures and light intensities largely explain the high productivity of seaweeds in these temperate waters. Mobilization, translocation and utilization of stored carbon for new growth is more obvious in a study on *Laminaria hyperborea* (Lüning 1969) in which

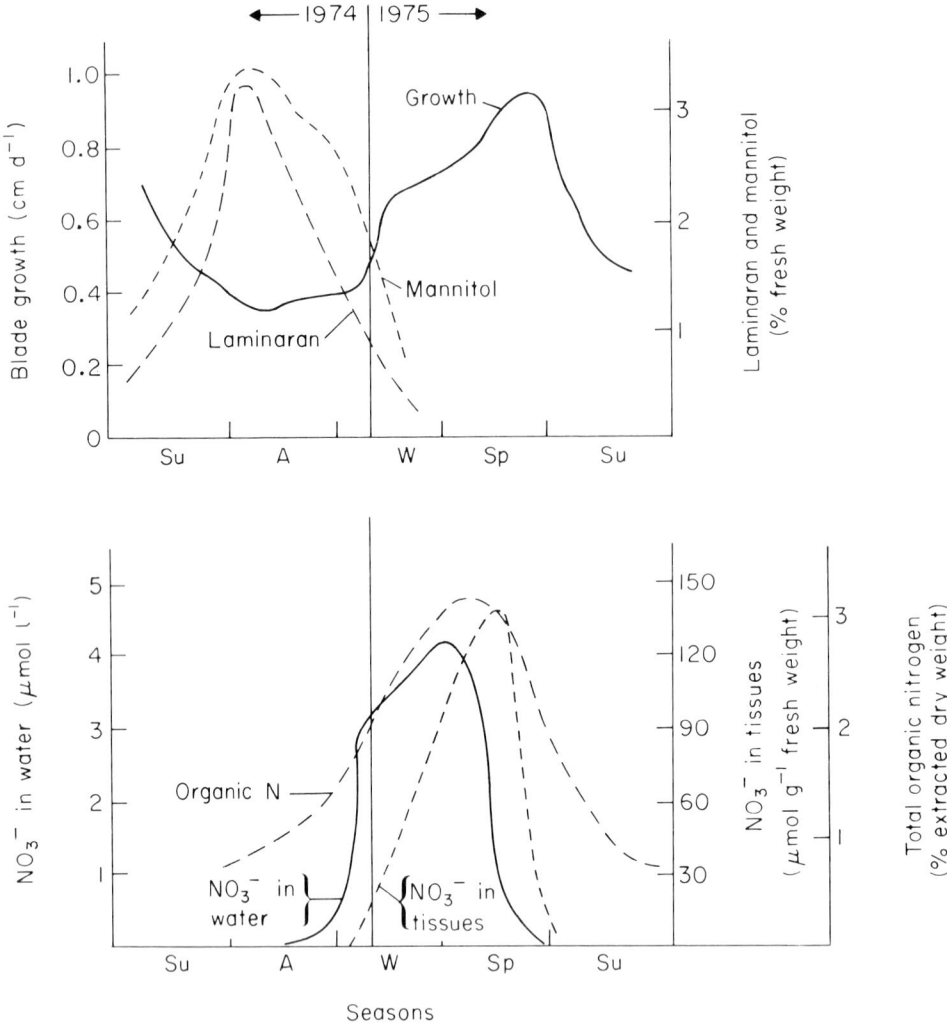

Fig. 6.3 Relationship between the rate of blade growth in *Laminaria longicruris*, its internal carbon and nitrogen reserves, and NO_3^- concentration in sea-water. NH_4^+ levels were highly variable between 0.2 and 2.0 μm l^{-1} with no apparent seasonal trend. (Smoothed curves from data at one 9 m depth station in Chapman & Craigie 1977, 1978.)

plants kept in the dark from January to June managed to produce a new frond.

Light

The photosynthesis versus irradiance responses of seaweeds appear to be basically similar to those reported in phytoplankton (see Fig. 3.4). There is a temperature-independent, initial slope saturating at a temperature-dependent P_{max} followed by inhibition at high light intensities (Arnold &

Murray 1980; King & Schramm 1976). The specific, light-saturated, photosynthetic rates (mg carbon fixed per mg algal carbon per unit time) in seaweeds are generally lower than those of planktonic algae and are related to the amount of non-photosynthetic, internal, structural tissues in the algal thallus. Thus, sheet-like and finely branched forms with little specialized structural tissue have higher photosynthetic rates than larger, thicker and coarsely branched forms. The more productive species tend to be annuals in the intertidal zone, whereas the subtidal, perennial species tend to have lower P_{max} rates.

Ramus and his colleagues (see Ramus et al. 1977) have conducted a series of short-term experiments on light adaptation, thereby avoiding the complications of seasonal effects. They suspended green, brown and red algae in the water column near the surface and near the compensation depth for seven days and then measured the changes in pigment content and photosynthetic performance. Since the experiments were carried out in a water column, both light intensity and light quality varied with depth. In all cases the concentrations of chlorophylls and accessory pigments increased (1.4–5 fold) during incubation in deeper water. In another set of observations, higher pigment concentrations were found in these same species growing in naturally shady spots in the intertidal zone compared with plants growing in naturally sunny spots. This inverse correlation of pigment concentrations and light intensity is also observed in phytoplankton (p. 27).

The ratios of accessory pigments to chlorophyll a showed varied responses. For the plants suspended in the water column, the most pronounced increases occurred in the phycoerythrin/chlorophyll a ratios in red algae with slight increases in chlorophyll b/a ratios in green algae. That these shifts in pigment ratios were not observed in plants growing in sunny and shady areas of the intertidal zone, suggests that light quality rather than intensity is the controlling factor. Thus, intertidal species showed intensity (sun–shade) adaptation in order to more effectively utilize lower light levels in shady locations, while algae suspended in the water column near their compensation depth exhibited intensity as well as chromatic adaptation in order to more effectively utilize the low levels of blue-green light found at depth. In intertidal brown algae, the actual concentration of all pigments also increased although the fucoxanthin/chlorophyll a ratio unexpectedly decreased with incubation depth. In summary, chromatic adaptation in seaweeds appears to be similar to that observed in phytoplankton in that algae containing phycobiliproteins show the greatest response. Although it is generally accepted and intuitively satisfying that red algae are well adapted to deep-water habitats by virtue of the presence of phycoerythrin, a recent analysis of photosynthetic action spectra suggests that light quality is not a major factor contributing to the success of red algae in deeper water (Dring 1981).

Photosynthetic rates in the shade-adapted seaweeds studied by Ramus et al. (1977) were higher on a freshweight basis than in the sun-adapted plants at all light intensities tested. These results are similar to the usual response brown algae examined. Uptake is generally stimulated by light as in phytoplankton, although in *Fucus*, no light stimulation was observed even after 20 h of darkness. The rate of NH_4^+ and NO_3^- uptake is enhanced approximately two fold in *Fucus distichus* following a period of partial (30–40%) desiccation (Thomas & Turpin 1980). Phosphate uptake is also stimulated by desiccation, but only in P-deficient plants. This enhancement presumably allows nutrient assimilation to catch up with carbon fixation which has continued during the period of low-tide exposure (see p. 99).

Symptoms of nitrogen deficiency in seaweeds are similar to those expressed in phytoplankton (p. 36). (Phosphorus deficiency has not been investigated in detail.) Nitrogen limitation results in lower growth rates, decreased content of nitrogen (both organic and inorganic), decreased chlorophyll content, decreased phycoerythrin content in red algae, lower light-saturated rates of photosynthesis and higher ammonium uptake rates. In *Gracilaria foliifera* the effect of light on the NH_4^+ uptake rate also depends on the nitrogen status of the alga (Fig. 6.6). Nitrogen-deficient plants

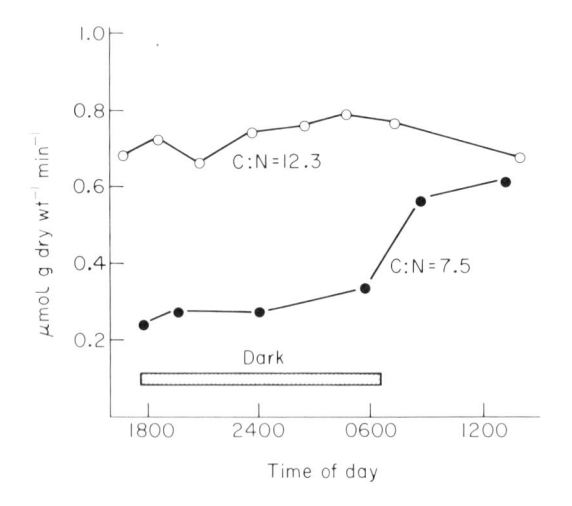

Fig. 6.6 Diel variation in the uptake rate of NH_4^+ at 20 μmol l^{-1} by *Gracilaria foliifera*: Nitrogen-sufficient plants (solid circle); nitrogen-deficient plants (open circle). (After D'Elia & DeBoer 1978.)

(C : N = 12.3) continue to take up NH_4^+ at high rates during the dark period whereas the dark uptake rate drops in nitrogen-sufficient plants (C : N = 7.5).

Temperature

Seaweeds exhibit the usual species-specific, metabolic responses to temperature that influence latitudinal distribution and seasonal patterns of growth. As one recent example, in a combination field and laboratory study, temperature was found to be a major factor influencing the seasonal growth of *Codium fragile*, a nuisance alga recently introduced to the north-eastern coast of the USA (Hanisak 1979). The interaction of temperature with light intensity and photoperiod in this organism is illustrated in Fig. 6.7. Higher

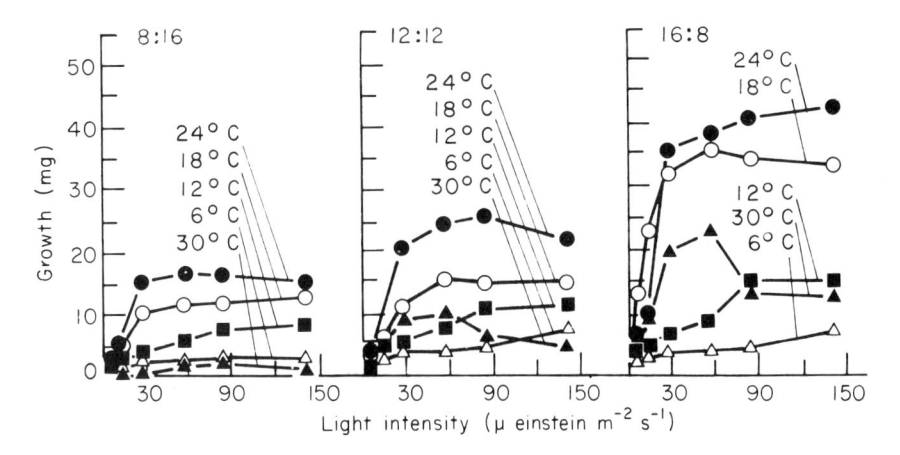

Fig. 6.7 Effect of temperature on the growth of *Codium fragile* at various light intensities in three different photoperiods. Growth was measured after three weeks. (After Hanisak 1979.)

light intensities are required to saturate growth at suboptimum growth temperatures (below 24°C). Inhibitory effects of high light intensities are more pronounced at high temperatures and longer photoperiods. Although growth was saturated at all temperatures within the range of light intensities tested, longer photoperiods resulted in increased growth. Seasonal adaptation in the temperature response of photosynthesis has been demonstrated in several seaweeds. In several red algae, for example, plants collected in the summer months have higher temperature optima for photosynthesis and are less sensitive to high temperatures than plants collected in the winter (Mathieson & Norall 1975).

Seaweeds in the intertidal zone often experience extreme fluctuations in temperature during exposure at low tide. Most subtidal algae, irrespective of geographical origin, do not survive freezing temperatures whereas the intertidal brown rockweed, *Fucus vesiculosus*, can survive several hours at $-45°C$. Tropical intertidal species cannot withstand freezing and some are even killed by exposure to temperatures below 5°C. Cold hardiness is not understood but may be related to an accumulation of organic solutes which lowers the freezing point and prevents ice-crystal formation. Partially drying plants before freezing enhances survival. High-temperature tolerance limits are generally about 27–35°C. Few species, even tropical ones, can survive 40°C. Again the mechanisms of heat hardiness are not understood, but are thought to be associated with resistance to enzyme denaturation. Drying also serves to improve heat tolerance; submerged specimens are much more sensitive to heat than are exposed plants.

Salinity

In many coastal habitats seaweeds will be subjected to periodic fluctuations in salinity associated with terrestrial run-off. In the intertidal zone, seaweeds will encounter severe salinity extremes: (1) as the algae dry out during low tide; (2) as water evaporates from high-shore rock pools between spring tides; and (3) when rain falls on seaweeds exposed at low tide. Just as temperature tolerance correlates with exposure, a number of studies have found that subtidal seaweeds will generally survive brief exposures to salinities of 15–45‰ whereas intertidal species may withstand salinites of 3–100‰. The tolerance range for growth is, of course, narrower than these values.

Like phytoplankton, seaweeds are able to adjust their internal osmotic potential by regulating the concentrations of inorganic ions and various organic molecules. Depending on the species, K^+, Na^+, or both K^+ and Na^+ are involved; Cl^- serves as the anion (Kirst & Bisson 1979). The organic molecules involved include mannitol in the brown algae and floridoside, isofloridoside or digeneaside in the red algae. Among the green seaweeds no characteristic molecules serving an osmoregulatory function have been identified. In contrast to most seaweeds which regulate cellular solutes to maintain a relatively constant turgor pressure, *Porphyra purpurea* simply tolerates adverse osmotic conditions (Reed et al. 1980). In dilute sea-water cell turgor increases and does not return to initial values during prolonged incubation. In hypersaline sea-water turgor pressure decreases and the cell shrinks in size.

Desiccation

In addition to temperature and salinity stress experienced during low-tide exposure, intertidal seaweeds also become dehydrated. Adaptation to desiccation varies with the vertical position in the littoral zone. Many upper intertidal species recover, following several days in the air-dry state, whereas low intertidal and subtidal species are killed by an exposure of a few hours or less. Relative humidity and temperature are obviously important variables in such experiments. The mechanisms of adaptation to drying conditions involve poorly understood, protoplasmic phenomena rather than any adaptation to retard the loss of water. The surface area/volume ratio of the plant is the principal factor governing the rate of water loss (Dromgoole 1980). Thicker-bodied forms lose water more slowly than thin, filamentous or membranous forms, with no significant difference between intertidal and subtidal species of similar morphology and no apparent correlation of morphology with the vertical position in the intertidal zone. Mucilaginous hydrocolloids (algin, fucoidan, agar, carrageenan) do not significantly retard water loss but could be involved in the ability of the algae to tolerate the desiccated condition. In the field, of course, features of the microhabitat (shady or protected areas and surrounding vegetation) will significantly affect the desiccation rate of individual plants.

In one of a series of papers on desiccation stress in intertidal fucoids, Schonbeck and Norton (1979) found variations in drought tolerance within a species. Specimens of *Fucus spiralis* collected from the upper part of its intertidal range tolerated a 48 h exposure which killed *F. spiralis* plants from the lower part of the range. Plants of both *F. spiralis* and *Pelvetia canaliculata* were less drought tolerant following transplantation for five and a half months to a lower intertidal zone and both species were less drought tolerant in the winter than in the summer. Physiological adaptation to desiccation stress was confirmed by laboratory experiments in which daily exposure of two to four hours for five days significantly increased drought tolerance in *F. spiralis*.

The preceding paragraphs may have left the general impression that exposure during low tide for most intertidal algae is a stress condition to be endured until the tide covers them once again and that adaptation to this habitat is one of tolerance rather than one of taking advantage of the situation to enhance growth. There is evidence, however, that some intertidal algae have higher rates of photosynthesis under emerged than submerged conditions (Quadir et al. 1979). Although rates of net photosynthesis immediately following emergence are lower than under submerged conditions, they increase during desiccation in algae collected from the upper intertidal zone. Maximum rates are observed at 20–40% water loss and may be over six times the submerged rates in some species. Photosynthesis declines with further desiccation, although these algae are able to maintain net carbon assimilation with up to 60–70% water loss, at least for short periods. In contrast, net photosynthetic rates in air for seaweeds from the lower intertidal zone are either equal to or lower than submerged rates and decline to zero after 30–40% water loss.

The molecular basis of these observations is not understood and it is not known if they are related to the apparent growth requirement in some seaweeds for periodic emersion (see below). It has been suggested that differences in the ability of upper and lower intertidal algae to use CO_2 and HCO_3^- as a substrate for photosynthesis could account for the variation in response to emerged conditions. The relatively rapid diffusion of CO_2 in air could explain enhanced photosynthetic rates in partially dehydrated, upper-intertidal algae. Most experiments on this problem have utilized gas exchange systems for measurements under exposed conditions and dissolved oxygen for submerged conditions. Direct comparisons might be made more easily if ^{14}C techniques were used for both conditions. In one such study (Kremer & Schmitz 1973), in which the plants apparently were not allowed to dry out, emerged and submerged rates were equal in *Fucus* species from the upper intertidal zone whereas emerged rates were lower in algae from the lower intertidal zone. The subtidal kelp, *Laminaria saccharina*, did not fix CO_2 in air.

Given the observations that intertidal seaweeds are well adapted to the rigours of their habitat and that there are well-defined, vertical zonation patterns in species distribution in the intertidal zone, one may ask if certain species actually require periodic exposure. Indeed, there are several reports in the literature that some seaweeds will not grow if continually submerged. More research is needed, however, to determine whether the exposure requirement is related to actual metabolic requirements or whether secondary effects of continual submergence (e.g. anoxia or bacterial growth) cause the plants to die. An experimental test of the influence of various emersion–submersion regimes can be carried out in the laboratory under defined conditions using tide-simulating devices. In a study of the growth of algal sporelings (Edwards 1977), there was little correlation between the vertical level of the seaweed in nature and the ability of the sporeling to grow at various degrees of exposure. Experiments were carried out at both 100% and 92% relative humidity. All 12 species grew best with 100% submergence but many also grew well when exposed 82% of the time (at 100% relative humidity). Exposure at 92% relative humidity resulted in a dramatic reduction in growth, even in species normally found in the upper intertidal zone (Fig. 6.8). This study supports the hypothesis that many of the

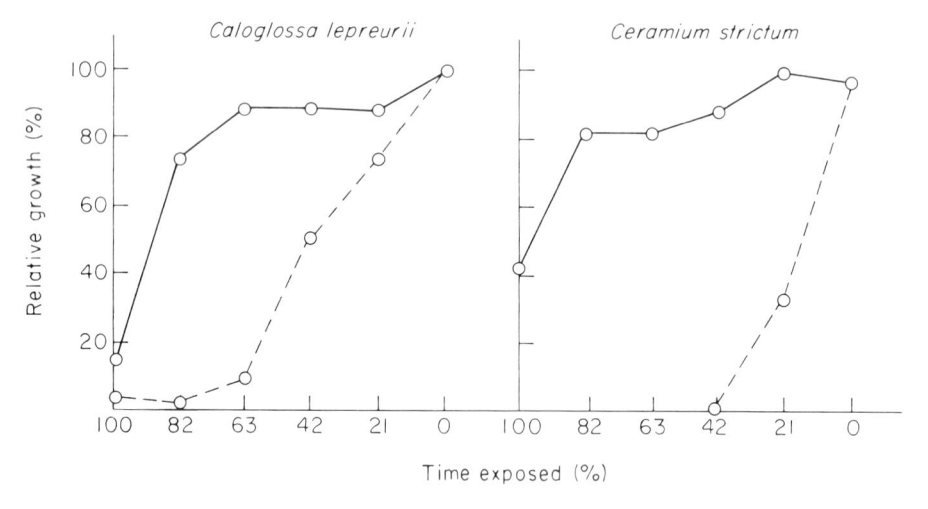

Fig. 6.8 Relative growth of sporelings after four to eight days in a range of tidal regimes from continual submergence to continual exposure. The percentage of time exposed to air of 100% relative humidity (solid line); and the percentage of time exposed to air of 92% relative humidity (dashed line). *Caloglossa lepreurii* is normally found in the upper intertidal zone while *Ceramium strictum* is normally found in the subtidal zone. (Data from Edwards 1977.)

algal species found in the intertidal zone live there because of competitive exclusion from more favourable habitats rather than because of any particular preference for the harsh conditions in the intertidal zone.

ENVIRONMENTAL CONTROL OF DEVELOPMENT

In unicellular organisms, growth and reproduction are almost synonymous since a doubling in biomass usually results in cell division to form two individuals. In multicellular species, however, the two processes are quite distinct and have different tolerance limits, those for reproduction being narrower than those for growth. Therefore, we must understand the environmental parameters that affect reproduction (e.g. formation and release of spores, establishment of spores on new substrates) as well as conditions that affect growth and morphogenesis, as part of the explanation for the success of a species in a given environment. Several examples are discussed below.

One useful technique for studying the effect of temperature and light intensity on growth and reproduction is a cross-gradient culture apparatus in which a temperature gradient is superimposed at right angles on a light gradient as shown diagrammatically in Figure 6.9. Each square represents a

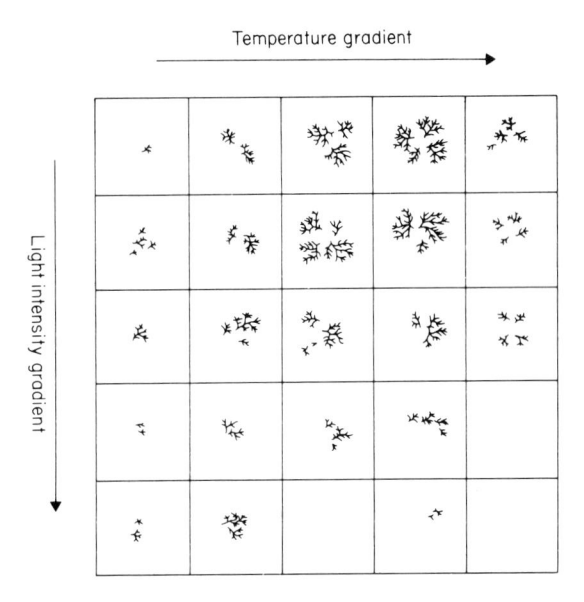

Fig. 6.9 Algal growth in a cross-gradient culture apparatus. Plants are grown at each combination of light intensity and temperature and then are mounted on herbarium paper and photographed.

culture chamber maintained at one of 25 different light and temperature combinations. The culture chambers are inoculated and the algae are allowed to grow for several weeks after which the plants can be mounted on a herbarium sheet and photographed for a dramatic visualization of the growth response. Additional notation may reflect the presence of reproductive structures at each light–temperature combination (Kapraun 1978). Alternatively, the plants may be examined in terms of their physiological responses to the two gradients.

The existence of the large, diploid, sporophytic phase in the life cycle of the kelps depends on environmentally controlled growth and gametogenesis in the microscopic, haploid, gametophytic phase (Lüning 1980). Haploid zoospores released in autumn or winter settle on the bottom and germinate. These gametophytes are well adapted to the low light intensities prevailing in temperate coastal waters during the winter as evidenced by growth saturation at very low light intensities. The gametophytes will even remain viable in the dark for several months. Gametogenesis in both male and female plants depends on a critical quantum dose of blue light rather than on a particular light intensity (quantum flux) or photoperiod. In the field this critical dose is accumulated over a number of days although in the laboratory it may be given in a single photoperiod. The dose of blue light required increases almost exponentially with increasing temperature. When mature, the egg is released from the oogonium and it releases a substance into the water that both triggers the release of spermatozoids and attracts them to the egg for fertilization.

There are several documented cases among the algae in which the life cycle is regulated by a change in the length of the day rather than by the total amount of light received (Cortel-Breeman & ten Hoopen 1978). This response, called photoperiodism, is well known in higher plants and animals and appears to be used as a means of predicting the seasons. Cold weather, for example, is forecast by shortening day length. The best-known case of photoperiodism in the algae is the short-day response in the life cycle of the sheet-like red alga, *Porphyra* (Dring 1970) which is extensively cultivated in Japan and eaten as 'nori'. This membranous plant alternates in the life cycle with a microscopic, filamentous, 'conchocelis' phase which inhabits calcareous shells and is the perennating phase in which the species survives the unfavourable season. The conchocelis phase of *Porphyra tenera* is induced to produce conchosporangia by the decreasing day length during the autumn. The liberated conchospores germinate to form the membranous phase which thrives and is harvested in the winter months, but disappears again in late spring. The critical day length required to produce conchosporangia is about 12.5 h or less. Seven or eight short-day (SD) cycles are required to fully induce the response although significant induction is achieved after two short-day cycles (followed by long-day cycles). A light break given during the dark period of an otherwise inducing SD cycle will inhibit conchosporangial development. Red light (660–670 nm) is most effective in causing the inhibition, while far-red light (730–750 nm) will reverse the effect of a previous exposure to red light. These responses are very similar to those of flowering plants and strongly suggest that the response in *Porphyra tenera* is mediated by the red, far-red, photoreversible pigment, phytochrome, found in higher plants.

Light gradients may also influence development in seaweeds. The zygote of the rockweeds (Fucales of the brown algae) develops the rhizoid on its dark side in order to attach the young germling to the substrate. The haptera of kelps grow away from the light, again an adaptation for attachment.

Water movements in the intertidal and subtidal environment are also involved in morphogenesis in some seaweeds. In the green seaweed, *Codium fragile*, the characteristic cylindrical axes of interwoven coenocytic filaments and the specialized utricles which form the compact surface layer of the axis only differentiate in the presence of shearing forces created by currents or, in the laboratory, by a reciprocating shaker. Without agitation, the filaments grow well but only produce a felty, green mass of prostrate and individual erect filaments (Ramus 1972).

Blade morphology in the kelps is also affected by currents and wave

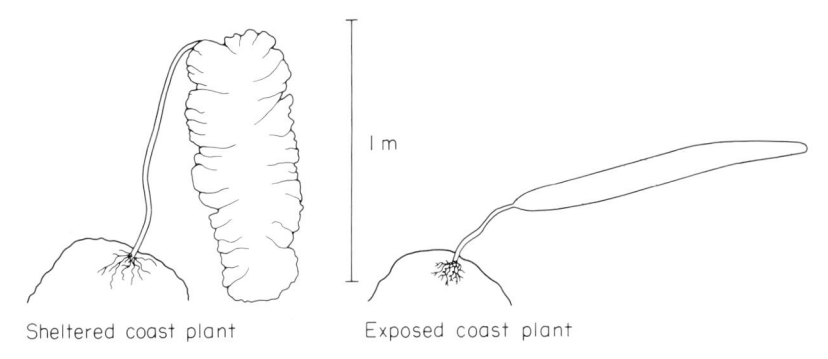

Fig. 6.10 Comparison of the morphology and posture of *Laminaria longicruris* plants growing in sites which are sheltered from or exposed to strong currents and swells. Plants in the sheltered site have long, hollow stipes and broad, thin blades whose surface-area/unit wet weight is three times that of the plants in the exposed site. The stipe is solid and shorter in the plants from the exposed site and the blades are three times as thick and one-fifth as wide as those of the sheltered plants. Blade length is similar at both sites. (After Gerard & Mann 1979.)

action. A recent study (Gerard & Mann 1979) comparing the growth of *Laminaria longicruris* from sites which differed in degree of exposure to wave action, points out the trade-off between beneficial and detrimental effects of water movement. As shown in Fig. 6.10, the morphology of the plants from the two localities differs markedly. One might predict that more intense water movement in the exposed site would enhance nutrient uptake to support higher growth rates than in the sheltered site during the nitrogen-limited summer months (discussed on p. 91). In fact the reverse was observed. The plants in the sheltered locality maintained higher growth rates into the summer months even though internal and external nitrogen supplies were similiar at the two sites. It was concluded that the greater surface area per unit biomass of the thin, wide blades of the sheltered plants more than compensated for the less intense water movement across the blades. The sheltered plants also replenished their internal nitrogen reserves more rapidly during the following winter than did the exposed plants. It appears that the possible benefits in terms of nutrient status at the exposed site were not realized because of the structural adaptation required to survive the intense wave action. The necessity for structural adaptation was demonstrated in transplant experiments in which plants from the sheltered site did not survive the first three-week sampling interval at the exposed site. In the reciprocal transplant experiment, new growth on plants from the exposed site exhibited morphological characteristics of sheltered plants within ten weeks. Until these transplants assumed the morphology of sheltered plants, they exhibited lower growth rates and lower replenishment rates of internal nitrogen reserves than the appropriate controls remaining *in situ* at the exposed site. These results further illustrate the importance of water movement to nutrient uptake in plants with the same morphology.

TRANSLOCATION

It is not surprising that kelps, a group which includes the largest algae, would possess some mechanism for distributing reduced carbon from sites of active photosynthesis to sites of rapid growth, but the remarkable similarity of the conducting cells to those of higher plants is unexpected (Sideman & Scheirer 1977). The most-advanced algal sieve elements (in *Macrocystis*) are comparable to those of higher plants in terms of sieve-plate pore size, the presence of callose and P-protein and degenerate, enucleate cytoplasm. In contrast to higher plants, algal sieve elements have cross-connecting sieve elements joining adjacent, longitudinal files of sieve elements; they are not associated with companion cells and they contain numerous mitochondria. Rates of translocation in the kelps (5–78 cm h^{-1}) are comparable to those in higher plants (20–250 cm h^{-1}), but the principal products translocated are mannitol and amino acids rather than sucrose as in higher plants.

Using ^{14}C tracer techniques, several studies (see Schmitz & Lobban 1976) have shown that transport during the growing season consistently occurs from the more mature portion of the blade to the growing regions of the plant, primarily the meristem and to a lesser extent the holdfast. In kelps with complicated developmental patterns (e.g. *Macrocystis*), the translocation pattern can be rather complex as several growing points compete for photosynthate. The similarity observed to date in the anatomy of translocation systems in higher plants and brown algae, leaves us looking forward to the time when the mechanisms of transport in these independently evolved systems are understood and can be compared (see Buggeln 1981).

ORGANIC EXUDATION

Seaweeds release dissolved organic matter in addition to losing particulate matter through erosion, grazing, etc. It is not clear why this release is termed exudation rather than excretion as it is for phytoplankton. Exuded material includes polysaccharides, complex nitrogenous materials and, at least in brown algae, phenolic compounds. As is the case for phytoplankton, there is considerable controversy over the amount released and whether or not healthy plants do it. It is likely that at least some of the released material is derived by solubilization of the extensive, cell-wall matrix polysaccharides in seaweeds.

Rates of exudation reported in the literature vary from undetectable to several thousand μg C g^{-1} dry weight h^{-1} and up to 30–40% of net carbon fixed. It is agreed that stress conditions, including those normally experienced during exposure at low tide result in significant losses of dissolved organic carbon when the plants are rehydrated. Part of the exuded material is likely to be of physiological origin since exudation rates are higher during photosynthesis than in the dark (Ragan & Jensen 1979). The use of ^{14}C to measure exudation rates could result in underestimates unless care is taken to allow the pools from which material is exuded to reach isotopic equilibrium (see p. 159). This problem would be significant if a large amount of the exuded material were derived from the cell-wall matrix.

An indirect approach has supported the high exudation rates determined by direct measurements. An annual carbon budget for *Laminaria longicruris* (Hatcher *et al.* 1977) failed to account for 35% of the net carbon assimilation after considering the inputs to new frond tissue (45%), stipe growth (8%) and storage of carbon in mature frond tissue (12%). The remaining 35% was assumed to be lost as dissolved organic carbon. This value is surprisingly high since this kelp is subtidal and not subjected to the intertidal stresses which are thought to accentuate exudation.

COMMUNITY INTERACTIONS

The vertical zonation of algae on rocky shores represents more than simply a sorting out of species by their relative tolerance to the exposure gradient. A number of recent studies in the temperate zone have described various mechanisms by which competition (seaweed–seaweed as well as animal–seaweed) and predation (on the algae themselves and on algal competitors) influence zonation and community structure. The rocky shore is a useful experimental site (compared with terrestrial habitats) for studying basic theories in community ecology because the species are relatively short lived, animals are attached and relatively immobile and are therefore easy to count and remove if desired, and gradients of physical stresses are readily available if one wishes to evaluate the effect of abiotic factors on biotic interactions.

In contrast to planktonic environments, space is the principal limiting resource in the rocky intertidal zone since to live there an organism must be firmly attached to the substrate. If a dense stand of seaweed develops, however, light becomes the critical factor as the canopy of algal fronds shades the rock surface and free space develops between the seaweed holdfasts. Dayton (1975) has divided intertidal algae into three functional categories. The larger forms which may eventually dominate the light resources are called canopy species and are often perennials. Obligate, understorey species are those which apparently require the shade and protection of the canopy species since they die out or die back to the holdfast when the canopy species are removed. Fugitive or opportunistic species are the intertidal weeds. They are often annuals with rapid growth rates which can quickly colonize newly opened clearings on the rock, only to be replaced by canopy species during ecological succession. Several ecologically important invertebrates complete the main cast of characters in this environment. Mussels are the dominant competitor in many habitats and are only held in check by their predators—certain snails and starfish. Herbivores seem to fall into two categories: the prudent molluscs, mainly limpets and periwinkles, which consume the fugitive species and nibble at larger forms and the seaurchins with voracious appetites which can decimate intertidal seaweed communities and kelp forests.

The following paragraphs describe several examples of community interactions on the rocky shore. These studies were carried out by experimentally manipulating or removing functionally important algae and invertebrates and then monitoring changes in community structure for several years.

Studies by Lubchenco (1978, 1980) illustrate several interesting interactions in the intertidal zone on the New England coast (Fig. 6.11). Starting

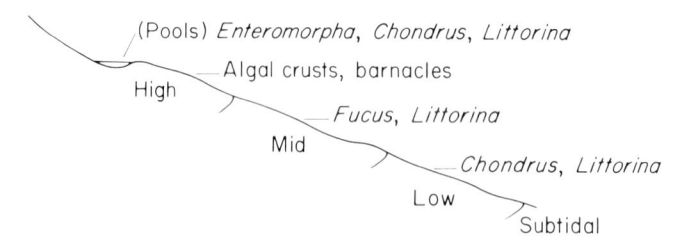

```
           (Pools) Enteromorpha, Chondrus, Littorina
           Algal crusts, barnacles
      High
                Fucus, Littorina
           Mid
                    Chondrus, Littorina
              Low
                     Subtidal
```

Fig. 6.11 Zonation of important organisms in the more protected areas of the rocky intertidal zone of the New England coast studied by Lubchenco (1978, 1980). In areas exposed to wave shock, mussels (*Mytilus edulis*) become the dominant organisms in the mid- and low intertidal zones.

the story in the tide pools, the most-important herbivore is a periwinkle, *Littorina littorea*, that prefers to eat the small and tender fugitive species (diatoms, *Enteromorpha*) rather than Irish moss, *Chondrus crispus*. The rapidly growing, fugitive species are able to outcompete slowly growing *Chondrus* in tide pools in which the *Littorina* density is low, but *Chondrus* dominates in those pools where *Littorina* density is high. Species diversity, therefore, is low at the extremes of periwinkle density, but at intermediate densities algal diversity is high because the periwinkles reduce, but do not eliminate, the fugitive algae. Pools dominated by *Enteromorpha* are not exploited by *Littorina* because there is little migration of the adult periwinkles between pools. Another stabilizing influence in the *Enteromorpha* pools is the presence of crabs which prey upon small periwinkles which are recruited from the plankton during their larval state. The crabs survive in the *Enteromorpha* pools because they can escape predation by sea-gulls by hiding beneath the *Enteromorpha* canopy.

In the mid-intertidal region (Fig. 6.11), the brown alga *Fucus* is often the dominant seaweed. Its upper limit is probably determined by physiological stress related to exposure. Experimental studies showed that its lower limit is determined by a competitive interaction with *Chondrus crispus*. When rock surfaces in the low intertidal zone are cleansed of all algae and their holdfasts (e.g. by sandblasting), the freshly cleared areas are colonized rapidly by *Fucus* (and fugitive, ephemeral algae) which is gradually replaced during the next several years by *Chondrus*. Even though *Fucus* has a high rate of recruitment and grows faster than *Chondrus*, it is the inferior competitor in this zone because *Chondrus* can maintain occupancy of space with its perennial holdfast. As the *Fucus* plants age and begin to die after two to three years, *Chondrus* slowly takes over the site. If only the *Chondrus* thallus is removed in these experiments, *Fucus* sporelings never become established on the *Chondrus* holdfasts. The upper limit of *Chondrus* is probably determined by its sensitivity to desiccation and its lower limit by sea-urchin grazing. In contrast to its effect in the high-tide pools, grazing activity by *Littorina littorea* in the low and mid-intertidal zones does not affect the outcome of the competition. *Littorina* actually speeds up the *Chondrus* take-over by removing the ephemeral algae and *Fucus* sporelings. Thus *Littorina* counteracts the 'normal' pattern of competitive dominance in the high-tide pools and reinforces it in the low intertidal zone.

In exposed areas of the New England coast subjected to wave shock, the ultimate competitive dominant for space in the low and mid-intertidal zones is the mussel, *Mytilus edulis*. Mussels are excluded from the more sheltered sites by their predators (starfish and a predatory snail) which cannot survive at the more exposed sites. In the absence of mussels, algae (*Fucus* or *Chondrus*) dominate the space. In summary, it is apparent that algal zonation in the intertidal zone of New England is determined by a complex set of interactions involving algal physiology, the physical environment and the presence and food preference of herbivores. Predation on various intertidal animals also may ultimately affect the seaweeds.

In the course of these studies additional observations were made concerning the adaptive significance of heteromorphic life histories in intertidal seaweeds. A number of species in the mid- to high intertidal zones exist as upright filaments, tubes or blades during the favourable seasons of the year and as crusts or boring stages during the physically harsher seasons. Although it had been proposed earlier that the non-upright morph functioned as a perennating stage during unfavourable seasons, studies in both New England and Oregon found that the upright morphs appeared during the unfavourable season in plots protected from grazers (Lubchenco & Cubit 1980). Thus, in at least some cases, heteromorphic life histories appear to be an adaptation to seasonal fluctuations in grazing pressure.

Along the Pacific coast of North America, sea urchins and their principal predators, sea otters, provide another interesting example of community interactions affecting seaweeds. Sea urchins are very important herbivores wherever they occur in significant numbers in subtidal coastal waters. Like periwinkles, they have strong food preferences which can result in the improvement of the competitive status of an unpalatable alga. In laboratory tests sea urchins were attracted chemically to their preferred food source (*Nereocystis luetkeana*) and grew rapidly when fed on it (Vadas 1977). The least-preferred algae tested (*Agarum* spp.) did not attract urchins and did not support as high growth rates. *Nereocystis luetkeana* is an annual which appears to have evolved an *r*-selected strategy (see p. 72) as a rapidly growing, fugitive species in order to survive predation by the urchins.

The near extinction of the sea otter by eighteenth and nineteenth century fur traders has resulted in large populations of sea urchins that have virtually denuded the subtidal zone of seaweeds in some localities. Reintroduction of sea otters is followed by declining sea-urchin populations and dramatic increases in subtidal kelps. In an experimental study in which urchins were removed manually (Duggins 1980), a dense canopy of *Nereocystis* appeared within one year and was largely replaced in the second year by *Laminaria groenlandica*, a perennial and the competitive dominant in this system. There is evidence that these dense kelp beds support increased fish populations which in turn provide the sea otters with an alternative food source as the preferred sea urchins become scarce.

As a final example, another kelp, *Egregia*, appears to suffer heavy damage from a limpet which grazes on the strap-like, blade-bearing rachis of the plant (Black 1976). The grazing scars frequently weaken the rachis to such an extent that it breaks and the plant loses a considerable amount of tissue. This loss did not affect growth rate or reproductive potential, but did stimulate the production of branch rachises. The results suggest that pruning by limpet grazing may actually improve the survival of the plant by preventing it from growing so large that it would be detached by wave action.

REFERENCES

ARNOLD K.E. & MURRAY S.N. (1980) Relationships between irradiance and photosynthesis for marine benthic green algae (Chlorophyta) of differing morphologies. *J. exp. mar. Biol. Ecol.*, **43**, 183–92.

BLACK R. (1976) The effects of grazing by the limpet, *Acmaea insessa*, on the kelp, *Egregia laevigata*, in the intertidal zone. *Ecology*, **57**, 265–77.

BUGGELN R.G. (1981) Source-sink relationships in the blade of *Alaria esculenta* (Laminariales, Phaeophyceae). *J. Phycol.* **17**, 102–4.

CHAPMAN A.R.O. (1979) *Biology of Seaweeds*. University Park Press, Baltimore.

CHAPMAN A.R.O. & CRAIGIE J.S. (1977) Seasonal growth in *Laminaria longicruris*: Relations with dissolved inorganic nutrients and internal reserves of nitrogen. *Mar. Biol.*, **40**, 197–205.

CHAPMAN A.R.O. & CRAIGIE J.S. (1978) Seasonal growth in *Laminaria longicruris*: Relations with reserve carbohydrate storage and production. *Mar. Biol.*, **46**, 209–13.

CORTEL-BREEMAN A.M. & TEN HOOPEN A. (1978) The short day response in *Acrosymphyton purpuriferum* (J. Ag.) Sjöst. (Rhodophyceae, Cryptonemiales). *Phycologia*, **17**, 125–32.

DAWSON E. Y. (1966) *Marine Botany: An Introduction*. Holt, Rinehart & Winston, New York.

DAYTON P.K. (1975) Experimental evaluation of ecological dominance in a rocky intertidal algal community. *Ecol. Monogr.*, **45**, 137–59.

D'ELIA C.F. & DEBOER J.A. (1978) Nutritional studies of two red algae. II. Kinetics of ammonium and nitrate uptake. *J. Phycol.*, **14**, 266–72.

DRING M.J. (1970) Photoperiodic effects in microorganisms. In Halldal P. (ed.), *Photobiology of Microorganisms*, pp. 345–68. Wiley Interscience, New York.

DRING M.J. (1981) Chromatic adaptation of photosynthesis in benthic marine algae: An examination of its ecological significance using a theoretical model. *Limnol. Oceanogr.*, **26**, 271–84.

DROMGOOLE F.I. (1980) Desiccation resistance of intertidal and subtidal algae. *Bot. Mar.*, **23**, 149–59.

DUGGINS D.O. (1980) Kelp beds and sea otters: An experimental approach. *Ecology*, **61**, 447–53.

EDWARDS P. (1977) An investigation of the vertical distribution of selected benthic marine algae with a tide-simulating apparatus. *J. Phycol.*, **13**, 62–8.
GERARD V.A. & MANN K.H. (1979) Growth and production of *Laminaria longicruris* (Phaeophyta) populations exposed to different intensities of water movement. *J. Phycol.*, **15**, 33–41.
HANISAK M.D. (1979) Growth patterns of *Codium fragile* ssp. *tomentosoides* in response to temperature, irradiance, salinity, and nitrogen source. *Mar. Biol.*, **50**, 319–32.
HATCHER B.G. (1977) An apparatus for measuring photosynthesis and respiration of intact large marine algae and comparison of results with those from experiments with tissue segments. *Mar. Biol.*, **43**, 381–5.
HATCHER B.G., CHAPMAN A.R.O. & MANN K.H. (1977) An annual carbon budget for the kelp *Laminaria longicruris*. *Mar. Biol.*, **44**, 85–96.
KAPRAUN D.F. (1978) Field and culture studies on growth and reproduction of *Callithamnion byssoides* (Rhodophyta, Ceramiales) in North Carolina. *J. Phycol.*, **14**, 21–4.
KING R.J. & SCHRAMM W. (1976) Photosynthetic rates of benthic marine algae in relation to light intensity and seasonal variations. *Mar. Biol.*, **37**, 215–22.
KIRST G.O. & BISSON M.A. (1979) Regulation of turgor pressure in marine algae. Ions and low molecular weight organic compounds. *Austral. J. plant Physiol.*, **6**, 539–56.
KREMER B.P. & SCHMITZ K. (1973) CO_2-Fixierung and Stofftransport in benthischen marinen Algen. IV. Zur ^{14}C-Assimilation einiger litoraler Braunalgen im submersen und emersen Zustand. *Z. Pflanzenphysiol.*, **68**, 357–63.
LUBCHENCO J. (1978) Plant species diversity in a marine intertidal community: Importance of herbivore food preference and algal competitive abilities. *Am. Natural.*, **112**, 23–39.
LUBCHENCO J. (1980) Algal zonation in the New England rocky intertidal community: An experimental analysis. *Ecology*, **61**, 333–44.
LUBCHENCO J. & CUBIT J. (1980) Heteromorphic life histories of certain marine algae as adaptations to variations in herbivory. *Ecology*, **61**, 676–87.
LÜNING K. (1969) Growth of amputated and dark-exposed individuals of the brown alga *Laminaria hyperborea.*, *Mar. Biol.*, **2**, 218–23.
LÜNING K. (1980) Critical levels of light and temperature regulating the gametogenesis of three *Laminaria* species (Phaeophyceae). *J. Phycol.*, **16**, 1–15.
MANN K.H. (1973) Seaweeds: Their productivity and strategy for growth. *Science*, **182**, 975–81.
MATHIESON A.C. & NORALL T.L. (1975) Physiological studies of subtidal red algae. *J. exp. mar. Biol. Ecol.*, **20**, 237–47.
NULTSCH W. & PFAU J. (1979) Occurrence and biological role of light-induced chromatophore displacements in seaweeds. *Mar. Biol.*, **51**, 77–82.
PAINE R.T., SLOCUM C.J. & DUGGINS D.O. (1979) Growth and longevity in the crustose red alga *Petrocelis middendorfii*. *Mar. Biol.*, **51**, 185–92.
QUADIR A., HARRISON P.J. & DEWREEDE R.E. (1979) The effects of emergence and submergence on the photosynthesis and respiration of marine macrophytes. *Phycologia*, **18**, 83–8.
RAGAN M.A. & JENSEN A. (1979) Quantitative studies on brown algal phenols. III. Light-mediated exudation of polyphenols from *Ascophyllum nodosum* (L.) Le Jol. *J. exp. mar. Biol. Ecol.*, **36**, 91–101.
RAMUS J. (1972) Differentiation of the green alga *Codium fragile*. *Am. J. Bot.*, **59**, 478–82.
RAMUS J. (1978) Seaweed anatomy and photosynthetic performance: The ecological significance of light guides, heterogeneous absorption and multiple scatter. *J. Phycol.*, **14**, 352–62.
RAMUS J., LEMONS F. & ZIMMERMAN C. (1977) Adaptation of light-harvesting pigments to downwelling light and the consequent photosynthetic performance of the eulittoral rock-weeds *Ascophyllum nodosum* and *Fucus vesiculosus*. *Mar. Biol.*, **42**, 293–303.
REED R.H., COLLINS J.C. & RUSSELL G. (1980) The effects of salinity upon cellular volume of the marine red alga *Porphyra purpurea* (Roth) C. Ag. *J. exp. Bot.*, **31**, 1521–37.
SCHMITZ K. & LOBBAN C.S. (1976) A survey of translocation in Laminariales (Phaeophyceae). *Mar. Biol.*, **36**, 207–16.

infection process. In these studies, laboratory-reared, algae-free (aposymbiotic) animals are challenged with cultured strains of symbiotic algae or with freshly isolated symbionts from various sources. Algae which are potential symbionts appear to be taken into the cell by phagocytosis as if they were food particles but they are not digested as a food particle would be. Recent work on *Hydra* suggests that antigenic determinants on the surface of *Chlorella* are responsible for the cell–cell recognition processes which result in the formation of a stable association (see Trench 1979) instead of the digestion or elimination of the cell.

The specificity of the host for the algal partner is not absolute. Although an infection often will occur with a number of related algal strains, certain strains achieve a more successful association than others. For example, several closely related genera of green flagellates will establish stable associations with *Convoluta*, but the worm grows best with its usual symbiont, *Platymonas convolutae*. This species also will completely replace other species which are established experimentally in the worms. *Chlorella* strains isolated from *Paramecium* can infect aposymbiotic *Hydra*. Strains of *Z. microadriatica*, isolated from various sources, vary widely in their ability to infect aposymbiotic specimens of a test species of sea anemone; some strains show no infectivity (Trench 1979). Except for *P. convolutae*, it is not clear whether or not the strains of similar algae should be recognized as separate species.

It is obvious that the growth rates of the symbiotic partners must be balanced so that one does not outgrow the other, but the regulatory processes involved are not understood. The number of algal cells per host cell varies from 1 to 300 in the various intracellular symbioses discussed here (Trench 1979). The variation is not as great within a single host and depends on the host species and the location of the cell within the host.

The exchange of metabolites between the symbiotic partners has been a popular research topic for a number of years. The selective advantage of the association is thought to be one in which the autotrophic partner feeds the heterotrophic partner and receives inorganic nutrients (waste products) in return. Oxygen released as a by-product of photosynthesis is not considered to be of significant benefit to the host in most cases because the organisms live in a well-oxygenated environment. Most studies have concentrated on the transfer of organic carbon from alga to host. Typically 20–50% of the carbon fixed by the alga is transported to the host, usually in the form of glycerol, glucose or maltose. Not as much research has been done on the transport of organic compounds or inorganic nutrients from host to alga. The early speculation that the animal 'farms' the algae and digests the excess algal cells has yet to be supported by strong evidence. It is possible that cells which appear to be undergoing digestion are actually scenescent algal cells undergoing autolytic degradation (Trench 1979). In fact, animals under stress (including dark starvation) usually expel the algae rather than digest them.

The invertebrate host often exhibits behavourial responses to light which appear to enhance photosynthesis by the symbiont. *Hydra* and *Convoluta* move toward the light (positive phototaxis) whether or not they are infected with their algal symbionts, suggesting that the control mechanism resides within the animal. In *Paramecium* only symbiotic individuals exhibit positive phototaxis; aposymbiotic cells are indifferent to light. Since the action spectrum for phototaxis corresponds to that of photosynthesis and since the response is only obtained under lower oxygen tensions, *Paramecium* may be responding to a photosynthetically derived O_2 gradient produced within the cell in a light gradient. A sea-anemone species which is capable of slow movement was found to move from the shaded to the lighted side of an aquarium in controlled experiments (Pearse 1974). Animals collected from deeply shaded natural habitats would move to the shaded area but would move back to the lighted area if the light intensity was reduced sufficiently. These experiments suggest that the sea anemone's behavourial response is adjusted as its symbionts undergo sun–shade adaptation. Aposymbiotic anemones, both naturally occurring and laboratory induced, did not show any phototactic response.

Platymonas and *Convoluta*

Convoluta roscoffensis is completely dependent on its symbiont since it will not grow or mature unless the larvae become infected. Its symbiont, *Playtomonas convolutae*, however, is found free living on the sandy beaches where the worm occurs. Once established in the worm, the alga loses its theca, flagella and eyespot. The worm may be grown through many generations solely at the expense of its symbionts, but in nature, it feeds during its young stages as the symbiosis is becoming established. If photosynthesis is blocked (by darkness or 3-(3,4-dichlorophenyl)-1,1-dimethylurea(DCMU), an inhibitor of photosynthesis), the mature worms apparently suffer some stress but survive for a month or more before dying of starvation. The appearance of lactic acid in these starved worms suggests that the lack of photosynthetic oxygen production has created anaerobic conditions in the worm. The accumulation of uric acid which occurs in dark-starved worms is thought to result from the cessation of algal growth. Cultured strains of *P. convolutae* can utilize uric acid as a sole nitrogen source for growth and presumably do so in the symbiotic association to recycle the nitrogenous waste products of the animal's metabolism. The *Platymonas–Convoluta* symbiosis is unusual in that glutamine and possibly other amino acids are the principal products released by the algae rather than carbohydrates as in other algae–invertebrate associations (Boyle & Smith 1975). These exchanges are illustrated in Fig. 7.1.

Chlorella and *Hydra*

Hydra viridis is much less dependent on its symbionts than is *Convoluta*. With an ample food supply in the light, green *Hydra* grows as fast as

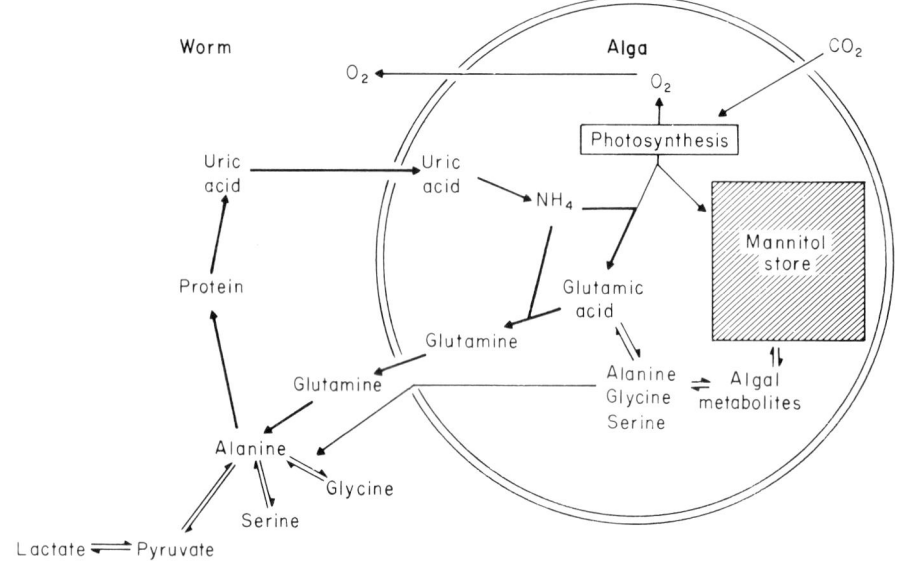

Fig. 7.1 Postulated pathways of nutrient interchanges in the symbiotic association between the green alga, *Platymonas convolutae*, and the flatworm, *Convoluta roscoffensis*. (After Boyle & Smith 1975.)

aposymbiotic *Hydra*, but the green individuals grow faster when food is limited. If forced to depend entirely on symbionts for food, green animals gradually decrease in size and survive about four weeks in the light while aposymbiotic animals only survive ten to twelve days. Aposymbiotic individuals are occasionally produced in culture and presumably also in nature, yet no colourless forms of this species are ever found in nature, suggesting that the supplemental nutrition provided by the algae is a selective advantage. The symbiotic *Chlorella* cells are morphologically indistinguishable from the well-known *Chlorella vulgaris* (which will not infect *Hydra*) and appear to undergo few if any morphological changes in symbiosis. Most of the carbon released to the animal is in the form of maltose. If symbiotic animals are fed and kept in the dark, the number of algal cells per animal remains relatively constant because the algae continue to divide and keep pace with *Hydra* reproduction. This observation suggests that the algae can live heterotrophically at the expense of food eaten by *Hydra*. Radioisotope experiments also suggest a transfer of sulphur and organic carbon from host to alga.

Chlorella and *Paramecium*

The *Chlorella* strains infecting *Paramecium bursaria* and *Hydra viridis* are very similar since they can cross-infect, although not always as effectively as the natural strains. The relationship is also similar in that the symbionts release maltose and increase *Paramecium* growth rates in under-fed but not in well-fed cultures. The high bacterial densities required for optimum growth are seldom encountered in nature suggesting that the algae provide a selective advantage for the protozoan. Again, aposymbiotic *P. bursaria* have not been reported in nature, even though *P. bursaria* eventually loses its symbionts when cultured in the dark. In contrast to *Hydra*, symbiotic *P. bursaria* can grow slowly in the light when depending solely on its symbiont in the absence of bacterial food.

Zooxanthella and Coelenterates

The symbiotic association between *Zooxanthella microadriatica* and various corals and sea anemones is the best studied of the algae–invertebrate symbioses. The relationship is not obligate since both partners can exist independently. As a symbiont the alga is a spherical, encysted cell with a reduced cell wall. In culture it assumes several forms including a non-motile, coccoid phase and the grooved, motile cell characteristic of dinoflagellates (Loeblich & Sherley 1979). Symbiotic animals kept in the light without an exogenous food supply remain healthier than aposymbiotic animals under the same conditions. In some cases the symbiotic association actually grows under these conditions, while in others, symbiotic individuals lose weight less rapidly than aposymbiotic animals. Differences are also observed in the animal's metabolism. A comparison of respiratory quotients in starved, symbiotic and aposymbiotic sea anemones (*Anthopleura elegantissima*) indicates that starved, symbiotic individuals (in the light) continue to catabolize carbohydrate provided by the zooxanthellae, while starved, aposymbiotic individuals rapidly begin to draw on lipid reserves (Fitt & Pardy 1981). Symbiotic animals kept in the dark gradually expel their algal symbionts over a period of a few weeks or months. Healthy animals in the light are also known to extrude pellets (distinct from fecal pellets) containing viable as well as degenerate algal cells. This process could be a means of controlling algal densities in the animal, but more definitive study is needed.

Several studies have found that zooxanthellae are capable of sun–shade adaptation while living as symbionts. For example, individuals of the reef-forming coral, *Stylophora pistillata*, growing near the surface at high light intensities are much lighter in colour than individuals from deeper water or from shady locations (Falkowski & Dubinsky 1981). The difference in pigmentation is due in part to four-fold, higher cellular concentrations of chlorophyll *a* in the zooxanthellae of the shade-adapted corals. There was no difference in the density of zooxanthellae per unit surface area of coral. P versus I curves (Fig. 7.2) show that shade-adapted corals have steeper initial slopes and reach P_{max} at lower light intensities, characteristics typically associated with shade adaptation. The relatively high compensation light intensities (where net O_2 exchange = 0) are due to respiration by coral tissue, but are also lower in shade-adapted corals. In both sun- and shade-adapted corals, the instantaneous light-saturated rate of photosynthesis by zooxanthellae exceeded the respiratory demands of the corals. No photoinhibi-

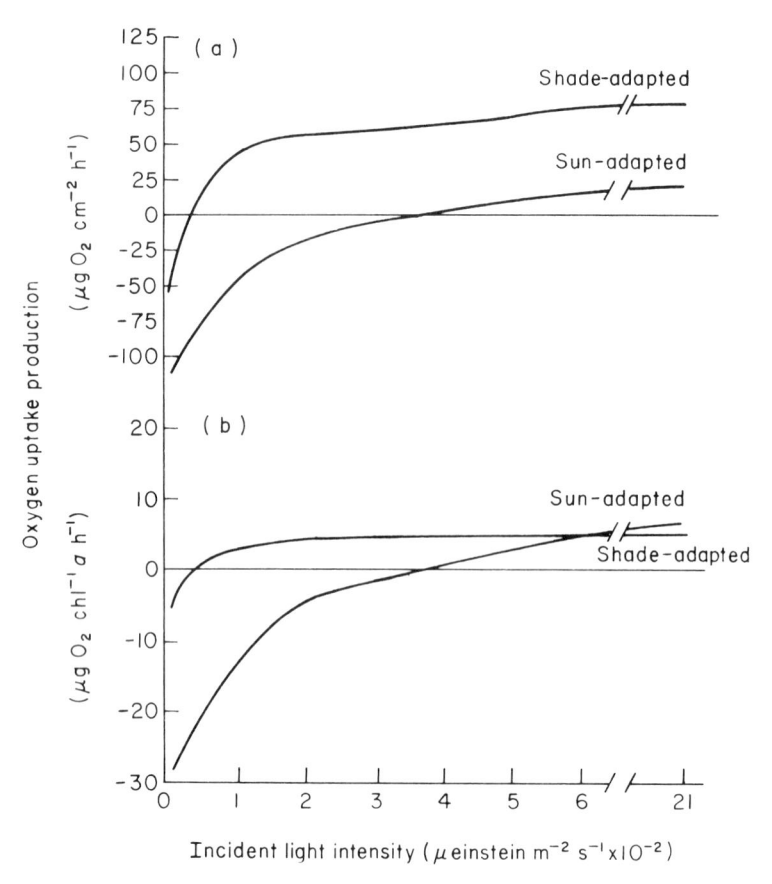

Fig. 7.2 Photosynthesis versus irradiance curves for shade-adapted and sun-adapted individuals of the coral, *Stylophora pistillata*. (a) Oxygen exchange per unit surface area of coral; (b) oxygen exchange per unit chlorophyll *a*. (After Falkowski & Dubinsky 1981.)

tion was observed. In transplant experiments zooxanthellae adapted characteristics consistent with the new light regime within four to eight weeks.

Translocation of photosynthetically derived organic carbon from algal symbionts to the host has been demonstrated in a number of different coelenterates (Muscatine 1973). In one type of experiment whole symbiotic animals are exposed to ^{14}C as HCO_3^- and then homogenized and fractionated to separate the intact algal cells from host tissue. Within 24 hours 20–50% of the total carbon fixed by the partnership is found in the animal tissue, mostly in the form of lipid (with virtually all of the activity in the glycerol moiety) and protein. Control animals incubated in the dark or aposymbiotic animals incubated in the light fix less than 10% as much carbon, demonstrating that the algae are responsible for the light-dependent fixation. In a second type of experiment, freshly isolated symbiotic algae which are exposed to radiolabeled HCO_3^- excrete 20–50% of the total carbon fixed to the medium. Most of the carbon is usually in the form of glycerol, but depending on the species, varying proportions also are found in glucose, alanine and various organic acids. That the composition of excreted carbon differs from labeled intracellular products provides evidence that the excretion is not merely the result of cell lysis. Maximum rates of excretion are obtained only in the presence of diluted homogenate from the host. The active compound(s) has not been identified, but it is only present in symbiotic animals. Specificity is not very great since host homogenate from symbiotic clams (*Tridacna*) will stimulate release in zooxanthellae isolated from coelenterates. Several hours after isolation from their host the algae lose their sensitivity to host homogenate and also begin to incorporate more ^{14}C label into insoluble products (proteins, etc.). These changes are reminiscent of a return to the free-living condition in which most of the fixed carbon contributes to algal growth rather than being served up to the host cell.

At least one pathway of internal carbon cycling between alga and host has been identified in a coral (Patton et al. 1977). The zooxanthellae import carbon (probably in the form of acetate) from the host and use it to synthesize fatty acids at the expense of photochemical energy (Fig. 7.3). The fatty acids and photosynthetically derived glycerol later appear as triglycerides and wax esters in lipid globules which are situated in the host cytoplasm near the symbiont cells.

In spite of the fact that translocation of organic matter between zooxanthellae and the host is well documented, its quantitative significance to the host's nutritional and caloric requirements has not yet been established (Muscabine & Porter 1977). The issue has been best studied in corals where the zooxanthellae may constitute up to half of the total biomass (as protein) of the partnership. Analysis of the stomach contents of reef corals confirms

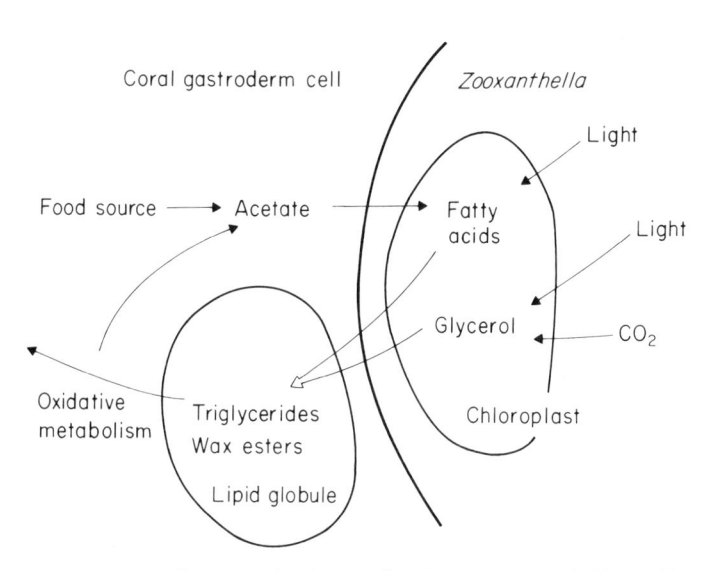

Fig. 7.3 Proposed scheme for internal carbon cycling between zooxanthellae and host cells in the coral *Porcillopora capitata*. (Modified from Patton et al. 1977.)

the impression that corals are unable to meet their daily requirements by feeding on zooplankton which are relatively scarce in the nutrient-poor tropical waters where coral reefs thrive. Further evidence is provided by the ratios of the stable carbon isotopes, ^{12}C and ^{13}C. The composition of the combined partnership is closer to that of the isolated zooxanthellae than to zooplankton captured in nearby waters, suggesting that most of the carbon in the coral tissue has passed through the algae. Using data in the literature, Muscatine and Porter (1977) have calculated that one species of reef coral is able to satisfy 87% of its daily carbon requirement for animal respiration from the carbon provided by the symbiotic algae. Even if it eventually is demonstrated that algae can provide enough organic carbon to meet the animal's needs for maintenance and growth, it is possible that some micronutrient requirements (vitamins, minerals) must still be met by feeding on zooplankton.

The success of symbiotic corals in nutrient-deficient waters has been attributed to an internal recycling of nitrogen and phosphorus between the heterotrophic host and the autotrophic zooxanthellae. Even if recycling were perfect, additional inputs of inorganic nutrients would be required to support additional growth. Inputs from both particulate food and from the dissolved nutrient pools in the water are possible. Recent research has demonstrated the ability of the partnership to take up dissolved nutrients and to retain (conserve) previously acquired nutrients (Muscatine & D'Elia 1978), but the actual interchanges between the partners remain to be identified. Corals with zooxanthellae are able to take up ammonium and phosphate from sea-water at natural concentrations; aposymbiotic corals continue to release these nutrients. Ammonium and phosphate uptake is stimulated by light. Ammonium uptake will continue in the dark for a number of hours, but after prolonged dark incubation (18 h), symbiotic corals release ammonia to the medium. Regardless of light conditions, corals release enough organic phosphate to the sea-water to offset the uptake of inorganic phosphate; the result is a net release of phosphorus from the symbiotic animal. It is assumed, but not yet demonstrated, that the zooxanthellae are the site of active uptake with the inorganic nutrients diffusing through the coral tissue to the algae. In any event, the demonstrated ability of the partnership to acquire and conserve nutrients must be regarded as an important adaptation for survival in the nutrient-poor tropical waters. Zooxanthellae therefore function to lessen the animal's dependence on holozoic feeding to supply reduced carbon and minerals.

Zooxanthellae also play an important role in the calcification process that builds the coral reef, although the mechanisms are not understood. The importance of algae is demonstrated by the observations that all reef-building corals contain zooxanthellae, that light stimulates the deposition of $CaCO_3$ and that an inhibitor of photosynthetic carbon fixation (DCMU) removes the stimulatory effect of light. It was originally proposed that algae stimulate calcification by removing carbon dioxide and shifting the following equilibrium to the right:

$$2\,HCO_3^- \rightleftharpoons CO_2 + CO_3^= + H_2O \qquad \text{(Eqn 7.1)}$$

It now appears more likely that the stimulatory effect of algae is indirect. It is possible that photosynthetically derived organic carbon provides an additional energy source which stimulates coral metabolism or that the additional carbon is used to synthesize the organic matrix component of the $CaCO_3$ skeleton.

Chloroplasts in sea slugs

Sacoglossan opisthobranch molluscs (sea slugs) feed on siphonaceous green algae by puncturing the cell wall and sucking out the contents. The lack of individual cell walls in these algae undoubtedly facilitates this type of feeding behaviour. Several slug species retain the chloroplasts within the cells of the digestive diverticula where they may remain structurally intact and photosynthetically functional for up to three months. These symbiotic chloroplasts contribute to the slug's metabolism much as symbiotic algal cells contribute to their host's well-being. Slugs starved in the light lose weight more slowly than slugs starved in the dark. Glucose appears to be the principal photosynthetic product transferred to the animal which uses it in mucus production and to meet respiratory needs.

The remarkable stability of the symbiotic chloroplasts away from the plant has raised questions concerning their biochemical autonomy (Trench & Ohlhorst 1976). Since it is well known that many chloroplast proteins required for chloroplast growth and division are coded for by nuclear genes in the plant cell, it is important to establish the extent to which the symbiotic chloroplasts are able to synthesize their own macromolecules. The available evidence supports the conclusion that the chloroplasts do not grow and divide in the slug. Although they synthesize small molecules, including amino acids and carotenoid pigments, and are capable of limited protein synthesis, they do not synthesize chlorophyll, chloroplast membrane proteins, ribulose bisphosphate carboxylase or nucleic acids. Chloroplast survival in the slugs therefore appears to be a function of protein stability rather than *de novo* synthesis. The chloroplasts must also, of course, be protected in some way from digestion by the animal cells.

LICHENS

Lichens are the most familiar and widespread of the symbioses involving algae. They grow on soil, rocks and tree trunks and are found in such diverse habitats as deserts, sea-shores, tropical rain forests, mountain peaks and polar regions. Thallus structure varies from crustose forms which are tightly appressed to the substrate, to leafy (foliose) forms and highly branched (fructicose) species. Reindeer moss (a fructicose lichen which grows to a height of 20–30 cm) occurs over vast areas of the arctic tundra and is an important food for reindeer and caribou.

The fungal component (mycobiont) constitutes more than 90% of the biomass of most lichens. Nearly 29% of all species of fungi are found as mycobionts in lichens (Smith 1978) and most of these belong to the Ascomycetes. The algal component (phycobiont) is much more restricted taxonomically since only about 30 genera of green and blue-green algae are found in lichens. The unicellular green alga, *Trebouxia*, is found in some 70% of all lichen species (Smith 1975). Approximately 10% of all lichen species contain blue-green phycobionts; the heterocyst-containing filament, *Nostoc*, is the most common form. Thallus morphology is determined largely by the mycobiont, although differentiation of the typical lichen thallus depends on the presence of the phycobiont. Cases are known in which the same lichen fungus will develop very different morphologies if associated with different phycobionts. A few lichens are known to contain both green and blue-green phycobionts in different parts of the same thallus. In these species the green alga is the principal phycobiont and the blue-green alga is restricted to localized portions of the thallus called cephalodia.

The internal tissue of most lichens is differentiated into three or four layers, with the algae localized in a layer close to the surface of the thallus but beneath a tightly packed cortex layer (Fig. 7.4). There is always very close contact between the fungal hyphae and the algal cells. In some cases the fungal hyphae form a deep invagination in the algal cell wall or actually penetrate the cell wall and cause an invagination of the plasma membrane. These haustoria-like penetrations are not as common as one might imagine and are not necessary for transfer of material between symbionts. Lichen reproduction may occur by fragmentation of the vegetative thallus or by the

Fig. 7.4 Cross-section of a typical foliose lichen showing the release of soredia. Algal cells in the thallus are localized in the upper part of the medulla. Rhizines are bundles of fungal hyphae that anchor the thallus to the substrate and conduct water by capillary action.

liberation of small, wind-disseminated fungus–algae units (called soredia) which may re-establish the lichen on fresh substrate. The mycobiont also regularly forms typical ascomycete fruiting bodies (ascocarps) which liberate ascospores. Since these spores must germinate in the vicinity of potential phycobionts to re-establish the symbiosis, their contribution to lichen reproduction is uncertain.

Both phycobiont and mycobiont may be grown separately in axenic culture. They show no unusual nutritional requirements compared with free-living relatives, but they tend to grow much more slowly. In culture the axenic mycobiont does not form a differentiated thallus. Resynthesis of the lichen from its component organisms requires several months of nutrient-deficient conditions and a moist but non-liquid environment. A recently published technique (Ahmadjian & Jacobs 1981) involves mixing cultured phycobiont cells with clumps of mycobiont hyphae and inoculating the mixture onto newly cleaved mica strips which have been soaked in the algal growth medium. The mica strips are incubated in a light–dark cycle in flasks as shown in Fig. 7.5. In the specific case studied, the mycobiont

Fig. 7.5 Culture system used to resynthesize lichens from separately cultured components. A mixture of algal cells and fungal hyphae is inoculated onto the mica strip. (After Ahmadjian & Jacobs 1981.)

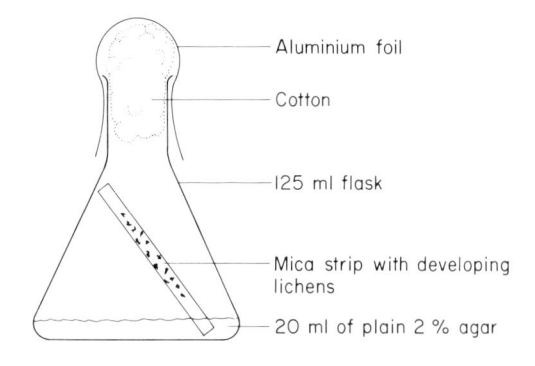

Cladonia cristatella lichenized several species of *Trebouxia* including its natural phycobiont, *T. erici*, but parasitized and destroyed a number of other species of green and blue-green algae isolated from various lichens. The nature of the envelopment of algal cells by fungal hyphae during the successful resynthesis process suggests that the relationship is more likely one of controlled parasitism than of mutualism.

As in algae–invertebrate symbioses, the algal symbiont in lichens releases a large amount of organic carbon (60–80% of the total carbon fixed) which is used by the heterotrophic partner. Blue-green phycobionts release glucose whereas green phycobionts release polyols such as ribitol, erythritol or sorbitol. The fungus converts the transported carbon to mannitol (and sometimes to arabitol as well) which accumulates to high levels in the fungus (Fig. 7.6). Release of organic carbon by the algae declines dramatically in a matter of hours or days following their isolation from the fungus. A search

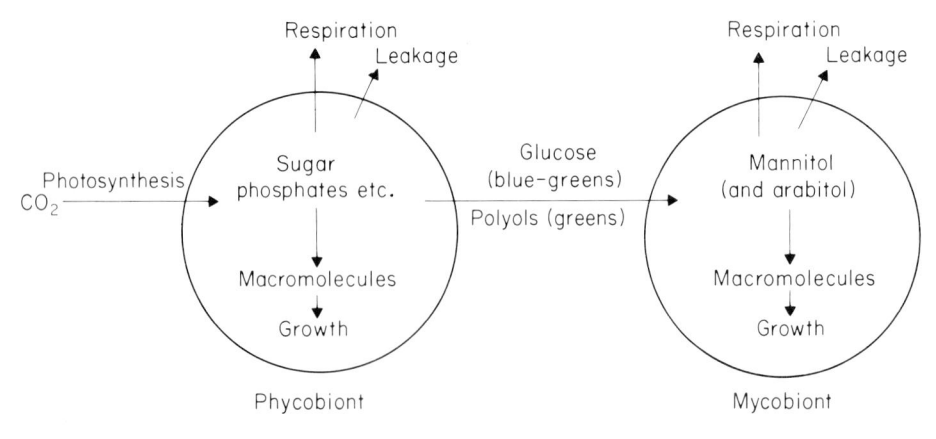

Fig. 7.6 Carbon flow through lichens.

has been made for chemical factors produced by the fungus which serve to stimulate carbon release by the algae. This approach follows the lead of results from algae–invertebrate symbiotic systems, but no evidence for the existence of such chemical factors has yet been uncovered. The release mechanism is currently explained as carrier-mediated, facilitated diffusion. The required concentration gradient is maintained by an accumulation of the product near the membrane in the alga and by active uptake by the fungus followed by conversion to mannitol.

The most significant problems in the physiological ecology of lichens relate to their ability to survive extreme and repeated desiccation and to grow in nutrient-poor environments. Both attributes are presumably directly related to their very slow growth rates. In contrast to most symbioses involving invertebrates, the heterotrophic partner in a lichen is completely dependent on the autotrophic partner as its source of organic carbon. The few available assimilation numbers (mg C fixed (mg chl a)$^{-1}$ h^{-1}) for lichen photosynthesis range from 0.49 to 1.92, values which are only slightly lower than those commonly encountered in phytoplankton (see p. 31). In spite of nearly normal photosynthetic potential, growth rates will be low since the alga, which is but a small fraction of the lichen biomass, has to support a relatively large amount of heterotrophic fungal tissue. Also contributing to the low net productivity of lichens is the obvious fact that they are only metabolically active when they are moist. This does not mean that they are dependent on rain to wet the thallus because a morning dew also will provide enough moisture. Lichens may also absorb enough water from the air, after several days at relative humidities above 80–90%, to resume metabolic activity.

Net productivity is also reduced by metabolic stress which occurs when an air-dry lichen thallus is suddenly rewet. Upon wetting, the respiration rate rapidly increases to several times the basal rate for a moist thallus and then slowly declines to the basal rate within a few hours. The function of this 'resaturation respiration', if any, is not known. Leakage of intracellular metabolites from both algal and fungal cells occurs for the first few minutes following a sudden rewetting, as the membranes become hydrated and return to their normal configuration (Farrar & Smith 1976). These losses to respiration and leakage become substantial with repeated cycles of hydration and dehydration. Thus, frequent experimental wetting of a lichen in its natural habitat would not necessarily increase growth because the lichen might not be able to make good the losses incurred during rewetting before it dried out again. It is not yet known how gentle rewetting (as by dew) compares with the rapid wetting of rain or laboratory treatments.

The polyol pool (mannitol and arabitol) in the mycobiont is thought to play a pivotal role in the lichen's ability to withstand desiccation stress. It acts as a 'physiological buffer' (Farrar & Smith 1976) by absorbing metabolic stress before the stress affects the growth and repair processes (proteins, etc.). Thus the losses to resaturation respiration and leakage come from the polyol pool with little change in the macromolecular components (Fig. 7.6). The turnover rate of photosynthetically acquired ^{14}C is much greater in the polyol pool than in insoluble components. By serving as a non-toxic osmoticum, the polyol pools, which are calculated to reach concentrations of several hundred millimolar, probably act to reduce water loss and may protect macromolecules from conformational changes during dehydration by direct substitution for water molecules in their hydration shells (Farrar & Smith 1976). Brock (1975) found that phycobiont photosynthesis per unit chlorophyll was less sensitive to low water potential (controlled osmotically or matrically) when the algae were in the lichen thallus than when they were isolated. The sparing effect of the fungus was ascribed to its polyol pool. This study provides some evidence to substantiate the often stated but seldom supported assumption that the alga is somehow 'protected' by the fungus. Since free-living algae are often found growing on or near lichen thalli, however, fungal protection would not appear to be an obligate requirement for survival in these habitats.

Since alternate wetting and drying imposes such severe stress on lichens, one would expect that lichens would thrive if kept moist. Most lichens do not. In order to maintain growing lichens in the laboratory, they must be subjected to periodic drying. Lichens maintained under saturating moisture conditions in constant light show a substantial decline in photosynthesis and phosphate uptake after one week (see Farrar & Smith 1976). Under this type of stress the polyol pool also exhibits the first symptoms of stress as it decreases in size. Under regimes involving alternate wetting and drying in the laboratory, the photosynthetic rate and polyol pools remained high although there was a great deal of carbon turnover due to losses during the wetting and drying.

Tysiaczny and Kershaw (1979) have provided an insight into the requirement for fluctuating moisture content in their study of the relative amount of recently fixed $^{14}CO_2$ which appears in mannitol. This fraction of the lichen was used as a measure of the fixed carbon which had been transported to the fungus. As shown in Fig. 7.7, the maximum transfer of

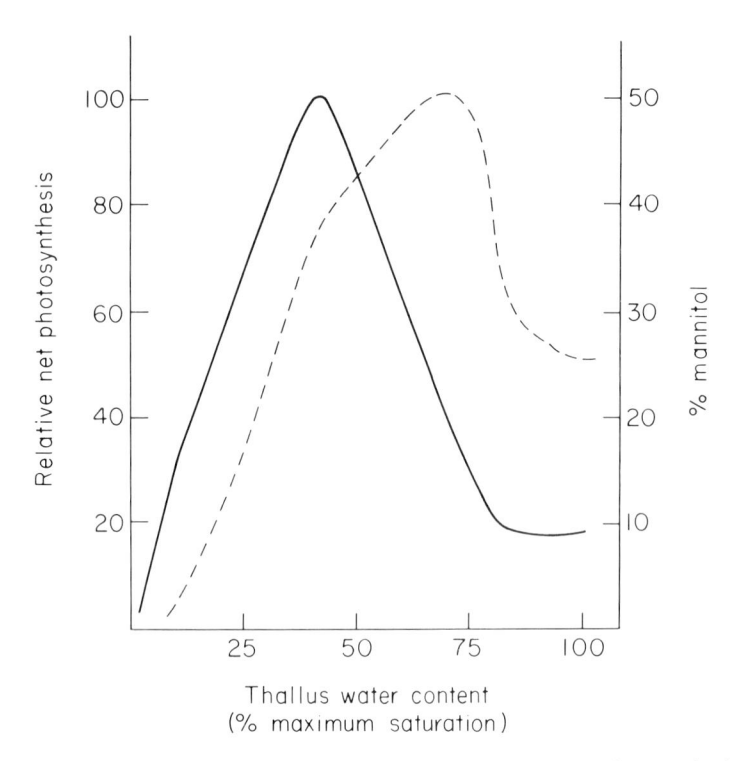

Fig. 7.7 Effect of thallus water content in the lichen *Peltigera canina* on photosynthesis and on the relative amount of carbon transported from phycobiont to mycobiont. Relative net photosynthesis per unit dry weight (solid line); ^{14}C-mannitol as the percentage of total soluble ^{14}C (broken line). (Data from Tysiaczny & Kershaw 1979.)

carbon does not occur at the same thallus moisture content as the photosynthetic maximum. Photosynthesis continues at low levels of thallus hydration at which point carbon transport (as measured by ^{14}C-mannitol) has become negligible. The alga appears to share its photosynthate with the fungus when the thallus is wet, but becomes progressively more stingy as the thallus dries out. Thus the proper distribution of photosynthate required to maintain the symbiosis depends on the fluctuating moisture status of the lichen. Parenthetically, it is interesting that the photosynthetic rate in this lichen is not maximal at 100% thallus saturation. In general, maximum photosynthetic rates in lichens are obtained between 50% and 90% saturation. A similar response has been observed in some intertidal seaweeds (p. 99).

The long-lived nature of lichens means that they must be able to survive the entire annual spectrum of environmental fluctuations while they are in the vegetative condition. Many freshwater and terrestrial algae and fungi are able to retreat into a resting spore of some type during unfavourable conditions. One of these adverse conditions is the temperature stress which often accompanies desiccation. As is the case in intertidal seaweeds (p. 97), lichens are more tolerant of temperature stress when they are dry, although not all species are as resistant to high temperatures as one might imagine. In at least one case (MacFarlane & Kershaw 1978), the temperature sensitivity of net photosynthesis and nitrogen fixation in two populations of *Peltigera canina* correlates very well with their habitat. In plants from a woodland site, these activities remained high (when thalli were rehydrated) in plants stored air dry at 25°C for 27–40 days, but declined by 80–100% in plants stored at 45°C. Plants from a drier roadside habitat were unaffected by storage at 45°C. In both populations respiration was not affected, suggesting that the phycobiont is more sensitive to temperature stress than the mycobiont. In another study (Kershaw 1977), two species of *Peltigera* were shown to have a remarkable capacity for temperature acclimation. The temperature optimum for net photosynthesis changed from 5°C in winter to about 30°C in mid-summer; the optimum winter rates were about one half the summer rates. There was little seasonal change in the response of photosynthesis to light intensity or thallus moisture content. During the early spring and late autumn, complete temperature acclimation occurred in either direction within 24 h in moist thalli. During winter and summer the capacity for temperature acclimation was much more limited. Respiration rates did not show any seasonal change suggesting that the alga rather than the fungus is the partner responsible for temperature acclimation.

In spite of the fact that lichens are typically found in nutrient-poor habitats, there is little evidence that they are actually nutrient-limited (Smith 1975). The reader will recall that in fact, low nutrient conditions are required to establish and maintain the symbiotic association. Nutrient levels in the thallus are similar to those in free-living algae and fungi. In nitrogen-poor habitats, there is no obvious selection for lichens that contain N_2-fixing, blue-green phycobionts. Nutrient-enrichment experiments have had mixed results (Armstrong 1977), enhancing growth rates in some cases but not in others. Therefore, it seems likely that rather than being caused by nutrient limitation, the slow growth rate of lichens actually requires only small amounts of nutrients. Rainfall may provide an adequate supply since it often contains μmol l^{-1} levels of combined nitrogen and somewhat lower levels of phosphate. The early part of the rain shower has a 'scouring' action in removing chemicals from the atmosphere and is therefore particularly rich in nutrients. Water streaming down tree trunks during a rain is also rich in elements such as calcium, potassium and magnesium.

This capacity to absorb elements from rain-water, coupled with the slow growth rate and longevity of lichens, results in the accumulation and concentration of certain elements within the lichen thallus. Differential sensitivity among lichen species to the toxic effects of these elements ultimately affects their distribution. Lichens are therefore useful indicators of atmospheric pollution. Many species are especially sensitive to sulphur dioxide.

Lichens containing heterocystous blue-green algae (usually *Nostoc*) fix N_2 at rates comparable to free-living forms. N_2 fixation is restricted to the heterocyst although it was once thought possible that fungal respiration could lower the O_2 tension in the thallus sufficiently to allow nitrogenase to function in vegetative cells. As in free-living forms, N_2 fixation in the lichen symbiosis is largely light dependent but continues slowly in the dark at ever declining rates as the internal carbon reserves are exhausted. The response of N_2 fixation to thallus moisture status is similar to that of photosynthesis except that there is often no decline at full hydration. Nitrogenase activity, like the photosynthetic apparatus, is able to survive long periods of desiccation. Even after 66 days of air-dry storage, full N_2-fixing activity in *Peltigera polydactyla* was restored within 12 hours (Kershaw & Dzikowski 1977).

The physiology of blue-green phycobionts is considerably modified relative to free-living blue-green algae. In symbiosis, nitrogenase is much less sensitive to inhibition by combined nitrogen (NH_4^+ and NO_3^-) than it is in free-living algae. Glutamine synthetase is normally found in the heterocyst of free-living forms and is responsible for converting NH_4^+ to glutamine which is then transported to the adjacent vegetative cells (see Fig. 4.5). In symbiosis, the activity of this enzyme in the phycobiont is very low and as a result almost all of the fixed nitrogen is excreted as NH_4^+ and is available for assimilation by the fungus. The scarcity of cyanophycin granules (see p. 63) in the vegetative algal cells suggests that the symbiotic blue-green is actually nitrogen depleted (Stewart & Rowell 1977). The release of NH_4^+ stops soon after the algal cells are isolated from the lichen. Thus the normal internal regulation by blue-greens of their N_2-fixing metabolism seems to disappear during lichenization by a fungus. The mycobiont presumably is involved in these transformations, but the mechanisms are not yet understood.

ANABAENA–AZOLLA SYMBIOSIS

Azolla is a genus of small, delicate ferns which float freely on the surface of ponds and sluggish streams. Numerous overlapping leaves develop along the horizontal stem; true roots growing at branch nodes hang down into the water (Fig. 7.8a). The leaves consist of a thin, colourless, lower lobe which floats on the water and a thicker, green, upper lobe which has a cavity containing the blue-green alga, *Anabaena azollae*. The cavity initially forms as a depression in the ventral side of the developing dorsal lobe (Fig. 7.8b) and is later sealed off by the growth of fern cells at the margin of the cavity (Fig. 7.8c). *Anabaena* filaments are common among the leaf primordia at the stem apex and presumably become trapped in the cavity at an early stage of development. The growth of the alga is slowed or possibly stops altogether long before it fills the cavity. Mucilage and a few fern epidermal hairs occupy the remainder of the cavity. Some of the hair cells have ultrastructural characteristics which suggest they may function in the exchange of metabolites between the alga and fern (Peters *et al.* 1978). Algal filaments are incorporated into the sexual stage (megasporocarp), thus insuring the continuity of the symbiosis into the next generation. This feature is important since *Anabaena azollae* is not known to occur in the free-living condition and *Azolla* populations usually experience abrupt declines after periods of vigorous vegetative reproduction.

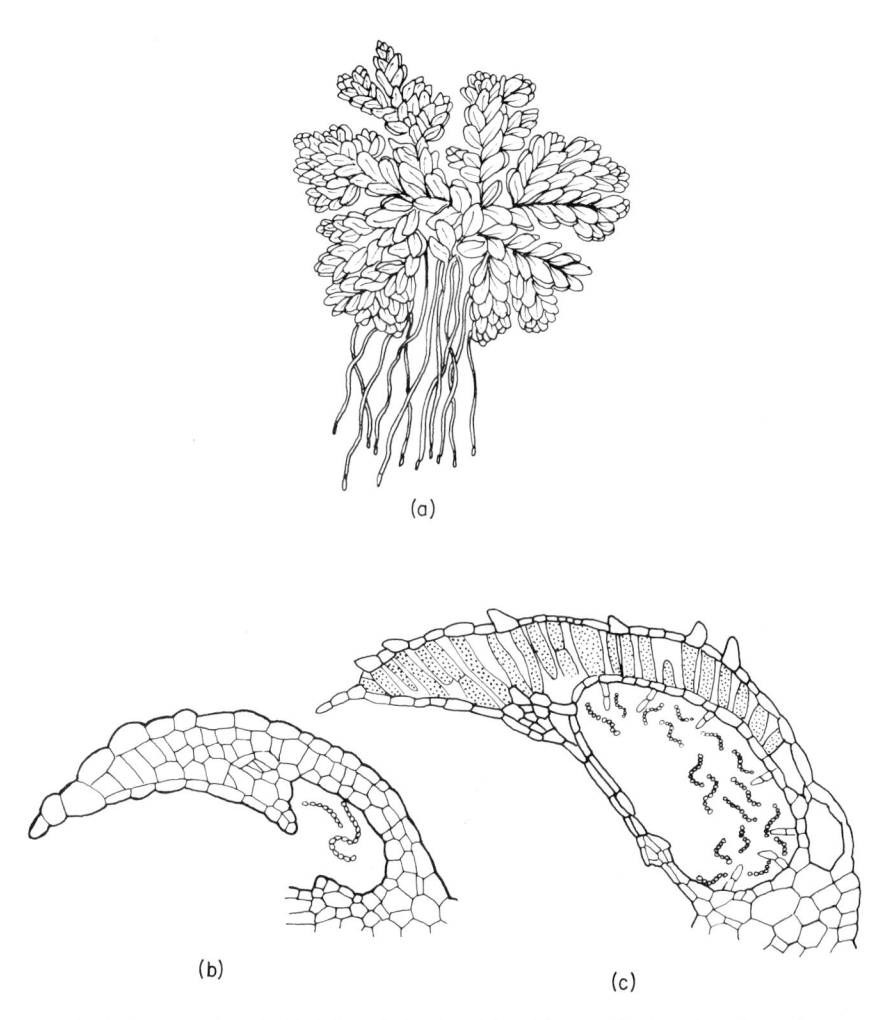

Fig. 7.8 The water fern *Azolla*. The whole plant (a) and longitudinal sections through an immature (b) and mature (c) dorsal leaf lobe. The N_2-fixing, blue-green alga *Anabaena azollae* is found in the leaf cavity. (b and c modified from Smith 1955.)

Azolla is able to meet all of its nitrogen requirements through the N_2-fixing activity of the symbiotic *Anabaena*, which accounts for approximately 15% of the biomass of the partnership. The alga's role is enhanced by a high heterocyst frequency (20–50%). Alga-free plants, which still develop the cavity in the dorsal leaf lobe, are occasionally found in nature and can be experimentally induced in the laboratory. They do not grow in the absence of combined nitrogen and they grow more slowly in the presence of combined nitrogen than do the symbiotic individuals in the absence of combined nitrogen. There are conflicting reports as to whether or not combined nitrogen enhances the growth of the symbiotic association. The N_2-fixing association is, however, unusually resistant to inhibition by NO_3^-. In free-living blue-green algae, nitrogenase activity is lost in a matter of days following the addition of NO_3^- to the medium, while N_2 fixation by the *Anabaena–Azolla* combination is only slightly inhibited after more than a month.

Although there are several reports that the blue-green symbiont has been cultured free of the fern (Newton & Herman 1979), the vast majority of the algal cells are apparently incapable of an independent existence. There is some doubt about the nature of the isolated strains because they are different in surface antigenicity (Gates *et al.* 1980) and will not reinfect alga-free plants. Perhaps the isolated strains represent mutants which are capable of independent growth. The physiological characteristics of the strains isolated from *Azolla* resemble free-living *Anabaena* species (most symbiotic algae revert to the free-living conditions when isolated), except that N_2 fixation under heterotrophic condition is two to three times higher than in free-living species (Newton & Herman 1979). This finding supports earlier suggestions that the fern may provide reduced carbon to support N_2 fixation by the alga. Freshly isolated algae release much of the fixed nitrogen as NH_4^+ and presumably do so in the leaf cavities of the fern.

Azolla is often considered an undesirable weed by virtue of its ability to rapidly form a multilayered mat on the water surface under favourable conditions. They may adversely affect the water column by restricting light penetration and O_2 diffusion. In south-east Asia, however, *Azolla* is a desirable organism which is used as a green manure in rice paddies. It is only useful under certain conditions, because the plant must die and decompose before its nitrogen is available to the crop. As fossil-fuel supplies dwindle and the cost of producing nitrogen fertilizer increases, scientists are trying to develop the *Anabaena–Azolla* symbiosis as one of our options for the future.

NOSTOC-BRYOPHYTE SYMBIOSIS

Several free-living species of the heterocystous blue-green alga *Nostoc* will form symbiotic associations with the liverwort *Blasia* and the hornwort *Anthoceros*. The relationship between the symbionts in these associations appears to be similar in many respects to that of the *Anabaena–Azolla* symbiosis, but the experiments are more definitive here because the organisms are more easily isolated from one another. Cavities on the underside of the prostrate gametophyte thallus become infected with compatible strains of *Nostoc* growing in the same habitat. Filamentous growths of bryophyte tissue invade the cavity but only in the presence of a *Nostoc* infection. These filaments presumably facilitate the transfer of metabolites between alga and bryophyte. The algal symbiont is not present in the sporophyte and is not directly passed on to the next generation through the sexual phases.

The *Nostoc* colonies (0.4–0.7 mm in diameter) are easily excised from the thallus for study. Cultures are readily established and may be used to reinfect algae-free individuals. Although the symbiosis is not species specific, successful associations were only achieved in algae-free *Blasia* by using *Nostoc* strains isolated from symbiotic bryophytes or from an angiosperm (*Gunnera*) which contains blue-green symbionts. A variety of other blue-green algae did not colonize the cavities.

Once the symbiosis is established, the heterocyst frequency in the *Nostoc* filaments increases from approximately 5% (in the free-living condition) to 30–50%. The remaining vegetative cells have negligible photosynthetic activity and very little, if any, phycobilin pigments. Rates of N_2 fixation increase due to the greater abundance of heterocysts. Isolated colonies liberate much of the fixed nitrogen in the form of NH_4^+, although they lose this capacity after three days of isolation from the bryophyte (Stewart & Rogers 1977). Studies on the interchange of nitrogen and carbon between *Nostoc* and the bryophyte are simplified in *Anthoceros* due to the presence of the alga-free sporophyte on the *Nostoc*-containing gametophyte. Tracer studies with ^{15}N have demonstrated the movement of fixed nitrogen from the alga through the gametophyte to the sporophyte within three days (Stewart & Rogers 1977). Although nitrogen-fixation rates in the dark are low and decline rapidly with continued dark incubation, sustained dark fixation rates were obtained by leaving the sporophyte in the light. The results suggest that carbon fixed by the sporophyte is transported to the alga to support nitrogen fixation in the dark. Using ^{14}C, direct evidence was obtained for carbon transport between an illuminated portion of the gametophyte and the *Nostoc* colonies in a darkened portion of the thallus.

The ease with which *Nostoc* cultures are obtained from this symbiotic association has prompted a comparison of free-living strains with the endophyte (Rogers 1978). Using N_2-fixing activity as an indicator of physiological health, various environmental conditions were investigated to determine if there were any conditions in which *Nostoc* benefited from the symbiotic association. Aside from receiving protection from the detrimental effect of low pH (pH 4.0–6.5), the endophyte does not appear to be in a more favourable habitat than the free-living form. In fact, N_2 fixation is more tolerant of desiccation in free-living *Nostoc* than in the endophyte. Since the alga grows slowly, if at all, once the cavity is completely colonized, it could be argued that the only benefits would be similar to those associated with forming a resting spore of some type. The symbiosis appears to be a one-sided exploitation of the N_2-fixing ability of the blue-green alga for the benefit of the host. This subject is considered in the following paragraph.

It is useful, if still somewhat speculative at this point, to summarize the available data to give a generalized view of the symbiotic relationship between heterocystous blue-green algae and eukaryotic photoautotrophs, namely those lichens which contain a second, green phycobiont, *Azolla*, and the symbiotic bryophytes (Stewart 1977). In the presence of the photosynthetic eukaryote to provide reduced carbon, the N_2-fixing capacity of the blue-green appears to be exploited at the expense of its CO_2-fixing capacity. There are several morphological and physiological differences between free-living and symbiotic blue-green algae which are apparently correlated with this exploitation (Stewart 1977). Thus, the heterocyst frequency increases from 3–5% to 20–60% with concomitant increases in N_2 fixation. The changes in nitrogen metabolism in the blue-green phycobiont which were discussed earlier (p. 126; loss of glutamine synthetase, excretion of NH_4^+, nitrogen-deficiency) are also evident. In at least one lichen, *Peltigera aphthosa*, glutamine (or a derivative thereof) produced by glutamine synthetase in the eukaryotic phycobiont (*Coccomyxa*) is involved in regulating nitrogenase activity in the *Nostoc* phycobiont (Rai et al. 1980). The blue-green phycobionts also are characterized by decreased photosynthetic rates, which helps explain, along with the nitrogen deficiency, why they grow so slowly as symbionts. The blue-green alga may even obtain a significant portion of its reduced carbon from its autotrophic associate. It is not known whether the eukaryotic symbiont produces some specific effector causing these changes in the algae, or whether internal regulation by the alga in the presence of the eukaryote is responsible (Stewart 1977). It is clear, however, that once free of the eukaryote this influence is lost and the blue-green alga reverts to the free-living state. *Anabaena azollae* is an interesting exception to the final generalization. By learning more about the relationships between N_2-fixing blue-green algae and their symbiotic partners (perhaps masters is a better word) it is possible that someday mankind will be able to manipulate blue-green algae in a similar way to produce nitrogen fertilizers with sunlight energy.

REFERENCES

AHMADJIAN V. & JACOBS J.B. (1981) Relationship between fungus and alga in the lichen *Cladonia cristatella* Tuck. *Nature (Lond.)*, **289**, 169–72.

ARMSTRONG R.A. (1977) The response of lichen growth to additions of distilled water, rainwater and water from a rock surface. *New Phytol.*, **79**, 373–6.

BOYLE J.E. & SMITH D.C. (1975) Biochemical interactions between the symbionts of *Convoluta roscoffensis*. *Proc. r. Soc. Lond. B.*, **189**, 121–35.

BROCK T.D. (1975) The effect of water potential on photosynthesis in whole lichens and in their liberated algal components. *Planta (Berl.)*, **124**, 13–23.

FALKOWSKI P.G. & DUBINSKY Z. (1981) Light–shade adaptation of *Stylophora pistillata*, a hermatypic coral from the Gulf of Eilat. *Nature (Lond.)*, **289**, 172–4.

FARRAR J.F. & SMITH D.C. (1976) Ecological physiology of the lichen *Hypogymnia physodes*. III. The importance of the rewetting phase. *New Phytol.*, **77**, 115–25.

FITT W.K. & PARDY R.L. (1981) Effects of starvation, and light and dark on the energy metabolism of symbiotic and aposymbiotic sea anemones, *Anthopleura elegantissima*. *Mar. Biol.*, **61**, 199–205.

GATES J.E., FISHER R.W., GOGGIN T.W. & AZROLAN N.I. (1980) Antigenic differences between *Anabaena azollae* fresh from the *Azolla* fern leaf cavity and free-living cyanobacteria. *Arch. Microbiol.*, **128**, 126–9.

KERSHAW K.A. (1977) Physiological–environmental interactions in lichens. III. The rate of net photosynthetic acclimation in *Peltigera canina* (L.) Willd var. *praetextata* (Floerke in Somm.) Hue, and *P. polydactyla* (Neck.) Hoffm. *New Phytol.*, **79**, 391–402.

KERSHAW K.A. & DZIKOWSKI P.A. (1977) Physiological–environmental interactions in lichens. VI. Nitrogenase activity in *Peltigera polydactyla* after a period of desiccation. *New Phytol.*, **79**, 417–21.

LOEBLICH A.R. III & SHERLEY J.L. (1979) Observations on the theca of the motile phase of free-living and symbiotic isolates of *Zooxanthella microadriatica* (Freudenthal) comb. nov. *J. mar. Biol. Ass. U.K.*, **59**, 195–205.

MacFarlane J.D. & Kershaw K.A. (1978) Thermal sensitivity in lichens. *Science*, **201**, 739–41.

Muscatine L. (1973) Nutrition of corals. In Jones O.A. & Endean R. (eds), *Biology and Geology of Coral Reefs*, Vol. II, Biology I, pp. 77–115. Academic Press, London.

Muscatine L. & D'Elia C.F. (1978) The uptake, retention, and release of ammonium by reef corals. *Limnol. Oceanogr.*, **23**, 725–34.

Muscatine L. & Porter J.W. (1977) Reef corals: Mutualistic symbioses adapted to nutrient-poor environments. *Bioscience*, **27**, 454–60.

Newton J.W. & Herman A.I. (1979) Isolation of cyanobacteria from the aquatic fern, *Azolla*. *Arch. Microbiol.*, **120**, 161–5.

Patton J.S., Abraham S. & Benson A.A. (1977) Lipogenesis in the intact coral *Porcillopora capitata* and its isolated zooxanthellae: Evidence for a light-driven carbon cycle between symbiont and host. *Mar. Biol.*, **44**, 235–47.

Pearse V.B. (1974) Modification of sea anemone behavior by symbiotic zooxanthellae: phototaxis. *Biol. Bull.*, **147**, 630–40.

Peters G.A., Toia R.E. Jr., Raveed D. & Levine N.J. (1978) The *Azolla–Anabaena azollae* relationship. VI. Morphological aspects of the association. *New Phytol.*, **80**, 583–93.

Rai A.N., Rowell P. & Stewart W.D.P. (1980) NH_4^+ assimilation and nitrogenase regulation in the lichen *Peltigera aphthosa* Willd. *New Phytol.*, **85**, 545–55.

Rodgers G.A. (1978) The effect of some external factors on nitrogenase activity in the free-living and endophytic *Nostoc* of the liverwort *Blasia pusilla*. *Physiol. Plant.*, **44**, 407–11.

Smith D.C. (1975) Symbiosis and the biology of lichenised fungi. *Symp. soc. exp. Biol.*, **29**, 373–405.

Smith D.C. (1978) What can lichens tell us about real fungi? *Mycologia*, **70**, 915–34.

Smith G.M. (1955) *Cryptogamic Botany. II. Bryophytes and Pteridophytes*. McGraw-Hill, New York.

Stewart W.D.P. (1977) A botanical ramble among the blue-green algae. *Br. phycol. J.*, **12**, 89–115.

Stewart W.D.P. & Rodgers G.A. (1977) The cyanophyte-hepatic symbiosis. II. Nitrogen fixation and the interchange of nitrogen and carbon. *New Phytol.*, **78**, 459–71.

Stewart W.D.P. & Rowell P. (1977) Modifications of nitrogen-fixing algae in lichen symbiosis. *Nature (Lond.)*, **265**, 371–2.

Trench R.K. (1979) The cell biology of plant–animal symbiosis. *Ann. Rev. plant Physiol.*, **30**, 485–531.

Trench R.K. & Ohlhorst S. (1976) The stability of chloroplasts from siphonaceous algae in symbiosis with sacoglossan molluscs. *New Phytol.*, **76**, 99–109.

Tysiaczny M.J. & Kershaw K.A. (1979) Physiological–environmental interactions in lichens. VII. The environmental control of glucose movement from alga to fungus in *Peltigera canina* v. *praetextata* Hue. *New Phytol.*, **83**, 137–46.

8 Algae in soil, snow and hot springs

The algae discussed so far in this book are the more obvious and abundant as well as the better-studied forms. A book on algal ecology, however, would be incomplete without a brief consideration of less frequently encountered algae. The discussion is brief because we know little of the physiological mechanisms by which these algae have adapted to what are often rather extreme and unlikely habitats. These algae may represent a valuable resource for experimental physiological ecologists, however. One useful approach to any physiological problem is to study the response under extreme conditions where the response is the most obvious and therefore more readily separated from other metabolic processes. By understanding how algae are adapted to unusual and extreme habitats, we can learn more about how the 'typical' alga is adapted to environmental fluctuations in its more moderate habitat.

SOIL ALGAE

Algae are virtually ubiquitous on or near the soil surface, even in soils which support few higher plants. Although under favourable circumstances a film or crust of algae may be visible to the naked eye, one usually must use culture techniques or microscopy to demonstrate their presence. Unicellular green algae, blue-green algae and diatoms are the most frequently encountered forms, although filamentous greens, yellow-greens and euglenoids are not uncommon. Blue-greens and diatoms are less abundant in acid soils than in neutral or basic soils, while the edaphic green algae appear to be more indifferent to soil pH. It has been suggested that soil algae may be ecologically significant, especially as pioneer species in the initial stages of succession. They may help to stabilize the soil surface against erosion, improve infiltration of water into the soil, improve soil texture by aggregating soil particles, add organic matter to the soil and help solubilize certain minerals. Their actual significance in these processes has not yet been extensively documented. It is clear, however, that nitrogen-fixing blue-green algae significantly improve the nitrogen status of soils, especially in rice production and other agricultural systems. The reader should consult Metting (1981) and Starks et al. (1981) for recent reviews on the biology of soil algae.

Much of the recent literature on soil algae has focused on taxonomic problems and floristic studies in various habitats (soils). Through the efforts of R. C. Starr, H. C. Bold and their collaborators, green algae which were once simply identified as *Chlorococcum* sp. or *Chlorococcum humicola* are now known to belong to a number of different genera. Identification of these unicellular, zoospore-producing algae requires careful study in culture using physiological and behavourial as well as morphological characteristics (Bold 1970). There does not appear to be a unique soil algal flora. Although certain genera (e.g. *Chlorococcum, Tetracystis*) are very common in soil and are rare in aquatic habitats, other genera (e.g. *Chlamydomonas* and *Characium*) are common in both. The small amount of information on seasonal patterns in the diversity of soil algae suggests that seasonal succession such as occurs in aquatic systems is not apparent in the soil (Broady 1979; Hunt et al. 1979).

The estimation of the algal standing crop is much more difficult in soil than in aquatic habitats. Fluorescence microscopy is a very useful improvement for direct microscopic counts but the technique is still tedious due to interference by soil particles. Estimates based on chlorophyll may be biased by even a small amount of vascular plant chlorophyll in the litter and humus. Culture-counting techniques such as agar plates or the serial-dilution (most probable number) method are often used, although they probably underestimate algal abundance due to cell aggregates (which would be counted as single cells) and the selective influence of the culture conditions and media. Estimates of 10^4 to 10^7 colony-forming units per gram of soil are commonly reported in the literature. Highest numbers are usually recorded at or within a few centimetres of the surface, although large numbers of living cells are reported much deeper in the soil, down to 1 m or more in some cases.

It is difficult to identify the important individual environmental parameters controlling edaphic algal biomass and productivity since so few experimental studies have been conducted. In the most complete study to date (Hunt et al. 1979), algal abundance was measured at monthly intervals (Fig. 8-1) and correlated with environmental conditions. Multiple regression analysis demonstrated that the abundance of blue-green algae was correlated with pH, season, Mg and precipitation; season and Mg were significant factors for eukaryotic algae. There was a positive interaction between the groups of algae with an increase in numbers of one group correlating with an increase in the abundance of the other. The seasonal effect was ascribed to fluctuations in light intensity rather than temperature. In this and other studies, the reduced light intensity beneath a forest canopy has been cited as the reason for lower, algal standing crops in such habitats. Low soil pH was probably responsible for the low number of blue-green algae at the forest site. Nitrogen and phosphorus do not appear to be as important controlling parameters as they are in aquatic environments. There are mixed reports on the effect of cultivation on the abundance of soil algae, causing increases in some cases and decreases in others. Soil com-

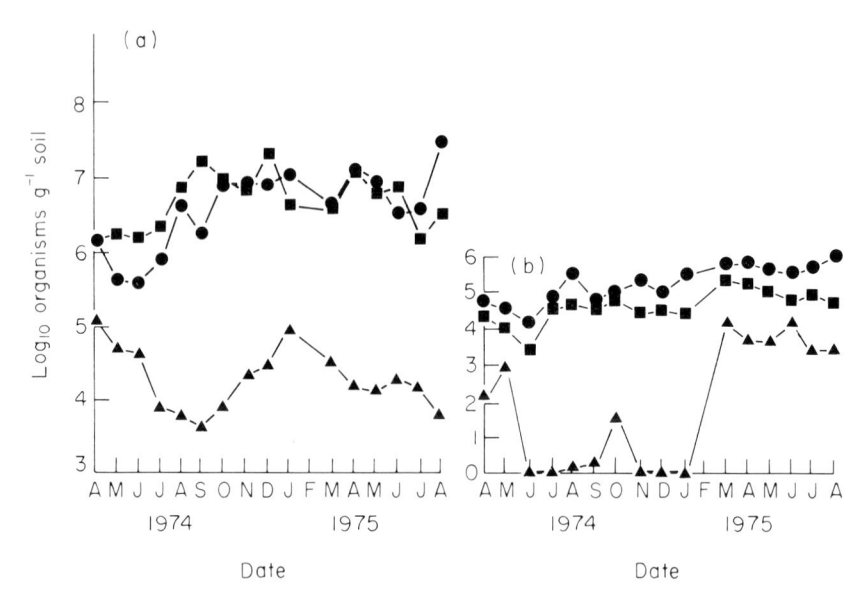

Fig. 8.1 Monthly changes in the number of soil algae in a one-year-old field (solid circle), an eleven-year-old field (solid square) and a forest (solid triangle). (a) Eukaryotic algae; (b) blue-green algae. (After Hunt et al. 1979.)

paction appears to increase the number of soil algae by some unexplained mechanism.

Although soil moisture would appear to be of overriding importance to soil algae, there has been surprisingly little research on the subject. Soil algae grow better in partially dry soil (40–60% of water-holding capacity) than in soil held at field capacity, reminding one of a similar response in some lichens and intertidal seaweeds. At the other end of the scale, soil algae may remain viable in some type of resting stage for 50–70 years in soils with a water content of 3–10% and up to 10 years in soils with a moisture content of 1% (air dry). Many soil algae, particularly green algae and blue-green algae are known to produce zygotes or akinetes which appear to be well adapted to withstand such extreme drought. There is no evidence that the thick cell walls on these resting cells are able to prevent water loss. Other soil algae, especially diatoms, are not known to produce such differentiated cells and apparently survive as vegetative cells which enter into some sort of physiological resting condition, often characterized by the accumulation of lipid and viscous protoplasm which is very resistant to plasmolysis. Although not yet investigated, it seems possible that polyol accumulation such as occurs in lichens could be one mechanism of drought tolerance in these algae. A number of studies have reported that these resting stages have increased resistance to extreme temperatures.

Cultured cells of a diatom were much more tolerant of desiccation stress when taken from old soil-water cultures than from young cultures or from cultures in defined medium at any age (Hostetter & Hoshaw 1970). Tolerance seemed to be associated with some unknown properties of the spent soil-water medium, since tolerance was lost within three hours if cells were transferred to fresh, defined medium. When combined nitrogen (the limiting nutrient) was added to old soil-water cultures, cell growth resumed, but the cells remained tolerant of desiccation stress.

In addition to not understanding the physiology of desiccation tolerance, we also know surprisingly little about what proportion of the viable units in the soil are in a resting stage at any one time or how long they require to resume activity once favourable conditions return. Are soil algae able to turn themselves on and off as quickly as lichens? Bristol Roach (1927) tested a technique that shows promise as a way to assay resting cell formation. By rapidly drying a soil sample in a stream of dry air at elevated temperature, she selectively killed the actively growing cells without damaging the resting cells. In her soil sample, assayed by the most probable number method, the drying treatment killed all but 12% of the cells present before the treatment.

The presence of living algae well below the euphotic zone (the top few millimetres) in the soil has prompted investigations into the heterotrophic capabilities of soil algae. While many soil algae are facultative heterotrophs in laboratory culture, experiments in soil kept in the dark show that algae grow very, very slowly or not at all, depending on the species. It seems likely that algae are washed into the soil during heavy rains and remain viable for many years in this moderated environment, perhaps utilizing the limited supply of usable organic matter in maintenance metabolism.

SNOW ALGAE

Although a snowbank would seem to be a very unlikely place to find algae, it is not uncommon to find them in bloom proportions, colouring the snow green, yellow, orange, pink or red (Hoham 1980). They appear in late spring or early summer when the air temperatures remain above freezing long enough to allow melt-water to form on the surface of the ice crystals. The vegetative cells can survive periodic freezing, but liquid water is required for spore germination and for the development of large populations. Most snow algae belong to one of a few genera of unicellular green algae (*Chlamydomonas, Chloromonas, Chlainomonas*) but a filamentous green algae, dinoflagellates, chrysophytes and cryptomonads have also been found growing on snow. The various non-green colours associated with snow chlorophytes are due to astaxanthin and other carotenoids which accumulate in vegetative cells and zygotes, probably as a photoprotective mechanism in response to the high light intensities in exposed snowfields. Nitrogen deficiency and normal maturation processes in zygotes may also contribute to carotenoid accumulation. In addition to the more obvious surface populations, snow algae may also be found in horizontal, green bands, associated with wetter, denser layers of snow (freeze–thaw bands), down to 40–50 cm in the snowbank.

The life histories of those snow algae studied in detail appear to be similar. Dormant zygotes spend the summer, autumn and winter in the soil (or on the old snow surface in persistent snowfields) where they undergo meiosis and cleave into daughter cells, a process that is stimulated by freezing temperatures. Germination in the spring apparently is triggered by melt-water and also may be influenced by as yet undefined conditions of light intensity, photoperiod, and/or dissolved nutrients and gases. The motile daughter cells swim towards the snow surface and rapidly develop into visible blooms. The vegetative phase lasts only a few days before gamete formation and fusion produce the next crop of zygotes. The motile zygote may remain motile for several weeks and continue to grow before maturing into a thick-walled zygospore. Although permanent snow is not required for the development of snow algae, each year's snow must be deep enough and persist long enough for the algae to complete their life cycle. Many early investigators, unaware of the short-lived vegetative phase or its relationship to the motile zygote and mature zygospore, described the various cell types they observed as separate species. Recent detailed field work by R.W. Hoham has demonstrated the relatedness of many of these forms.

Most species of snow algae are not known from other habitats, suggesting that they are well adapted to the rigours of living in snow. Culture studies of several species have demonstrated temperature optima for growth below 10°C. Other species which do not grow well in culture shed their flagella between 4°C and 10°C suggesting they also have temperature optima near freezing. The temperature response of photosynthesis in *Chlamydomonas nivalis* varied in different populations but usually exhibited optima at 10°C or 20°C with substantial activity at 0°C. No photoinhibition was observed at high light intensities by this species which is characteristically found in exposed areas. Another species, *Chloromonas pichinchae*, which is found in shaded areas, exhibits a well-defined phototactic response to avoid higher light intensities. The cells migrate down to 10–15 cm beneath the snow surface around dawn and return to the surface at dusk. The 1% light level in a snowbank can be as deep as 110 cm. It seems likely that snow algae would suffer nutrient deficiency, especially when concentrated in a layer on the snow surface. Nutrients leached from atmospheric dust and tree litter no doubt augment whatever nutrients occur in fresh snow. Extracts of coniferous leaf litter and bark stimulated the growth of one snow alga in axenic culture. It is difficult to imagine nutrient concentrations in the surface melt-water high enough to support visible algal blooms. It seems more likely that the high densities of non-motile cells become concentrated on the surface as the snow melts.

HOT-SPRINGS ALGAE

Moving to the other end of the temperature spectrum in aquatic systems, we also encounter conspicuous growths of algae in hot springs and their drainage channels. Hot springs are characterized by several interesting environmental conditions in addition to the obvious high temperatures. The absence of nearby vascular plants combined with shallow, clear water results in high light intensities. Total dissolved solids in hot-spring water are typically 10–15 times higher than in other surface waters and may vary considerably in composition in different springs. Some elements will be present at potentially toxic concentrations. Thus the few algae which have adapted to the thermal environment must be able to grow in a wide range of water chemistries if they are to colonize the available thermal habitats. Acidity, however, is one important restrictive factor. Blue-green algae are abundant in most hot springs above pH 6 but are completely absent below pH 4 and are rarely found below pH 5. Most blue-green algae are also sensitive to soluble sulphide, a frequent component of hot-spring water. In those few species of hot-spring blue-green algae able to tolerate sulphide, there is evidence that they may actually use $S^=$ as an electron donor for anoxygenic photosynthesis (Castenholz 1977). A similar process has been studied in more detail in planktonic blue-green algae (p. 63). In contrast to most other habitats which experience seasonal environmental fluctuations, the temperature and chemistry of any one hot spring are remarkably constant. The temperature gradient along and across the outflow channel is also fairly constant (if somewhat steeper in winter months) allowing the organisms to remain at their optimum temperature all year round. The drainways also provide convenient field incubators for experiments at a variety of temperatures. Most of the work on hot-spring algae and other organisms by Brock and Castenholz and their students has been summarized recently (Brock 1978).

It is now generally accepted that the upper temperature limit for prokaryotic organisms is higher than that for eukaryotic organisms and that heterotrophic forms tend to have higher temperature limits than photosynthetic forms. Thus some heterotrophic bacteria are capable of growing in boiling water but the highest temperature at which blue-green algae will grow is 70–72°C. The upper temperature limit for eukaryotic algae is 55–57°C (for *Cyanidium caldarium*) whereas some fungi can grow at 60–61°C. The molecular key to thermophily appears to reside in the lipid composition of the plasma membrane and other cellular membranes. Thermophiles tend to contain lipids with higher melting points (saturated instead of unsaturated fatty acids, for example) and as a result their membranes are more stable at high temperatures. Proteins from thermophilic organisms also exhibit increased stability, a useful observation that has not been overlooked by biochemists. There is nothing unusual about the ultrastructure of thermophilic organisms.

Blue-green algae dominate the flora in neutral and alkaline hot springs, usually growing in mats up to several centimetres thick. In most cases the algae only occupy the uppermost layer which may vary in colour from the typical blue-green to yellow, green or dark brown. The pink, red, orange or pale yellow, multilayered undermat consists of various autotrophic and heterotrophic bacteria. In outflow channels with a relatively stable temperature gradient (both downstream and from the centre to the edges of the channel), the blue-green algae form distinctive banding patterns as the various species occupy zones of optimum temperature.

Synechococcus lividus is a unicellular, rod-shaped blue-green that often occupies the high temperature zone in outflow channels. Several genetically stable temperature strains have been isolated which have temperature optima for growth at 45°C, 50°C, 55°C and 65°C. In their natural habitat these strains grow at slightly higher temperatures but in the same thermal sequence. In field studies the temperature optimum for photosynthesis of several populations along the thermal gradient was remarkably similar to the temperature at which the population was growing. Thus, these algae are actually adapted to these high temperatures and are not simply tolerant forms which are forced to live there because of competition at lower temperatures. The high-temperature form of *S. lividus*, in fact, will not grow below 54°C.

Synechococcus lividus is also capable of physiological adaptation to variations in temperature and light intensity. Cultures of the high-temperature strain were grown at 57°C and 70°C and then tested for their ability to photosynthesize at 71°C (Fig. 8.2). The culture grown at 70°C showed no adverse effects whereas the culture grown at 57°C lost all photosynthetic capacity within one hour at 71°C. Field studies on light adaptation have been conducted by shading portions of an algal mat for up to two weeks (Madigan & Brock 1977). P versus I curves for cell suspensions prepared from the top layer of unshaded and shade-adapted mats are shown in Fig. 8.3. Although some inhibition by full sunlight is evident in unshaded cells,

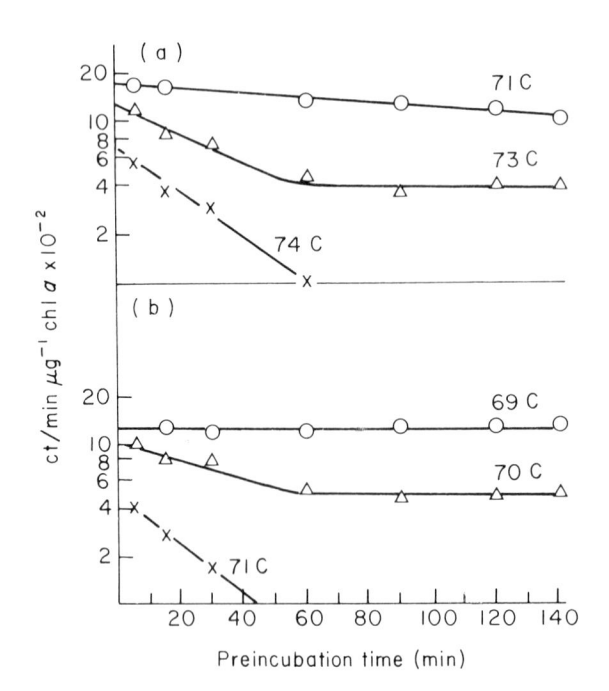

Fig. 8.2 Photosynthesis at supraoptimal temperatures by the high-temperature strain of *Synechococcus lividus*. Points correspond to incubation time at the indicated temperature prior to measurement of photosynthetic activity. (a) Cells cultured at 70°C; (b) cells cultured at 57°C. (After Meeks & Castenholz 1971.)

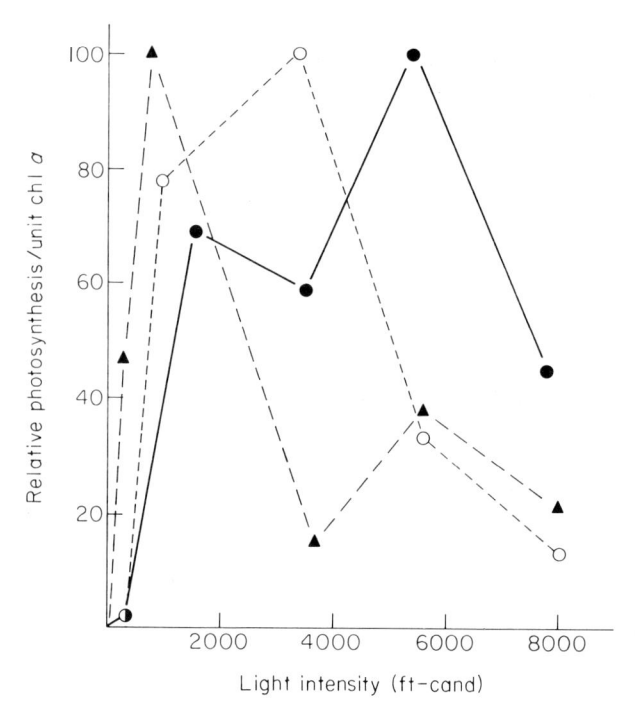

Fig. 8.3 Photosynthesis versus irradiance curves for suspensions of natural populations of *Synechococcus* growing at full sunlight (solid circle) or exposed to reduced light intensities (27% of full sunlight) for 10 days (open circle) and 16 days (solid triangle). (After Madigan & Brock 1977.)

shade adaptation results in a decrease in the optimum light intensity for photosynthesis and increased sensitivity to high light. Shade adaptation was accompanied by an increase in phycocyanin content while no significant changes were found in chlorophyll *a* content. In these experiments, the photosynthetic inhibitor DCMU (3-(3,4-dichlorophenyl)-1, 1-dimethyl urea) was used to distinguish blue-green photosynthesis from bacterial photosynthesis. Algal photosynthesis was calculated by subtracting DCMU-resistant photosynthesis (by the bacterium, *Chloroflexus*) from total incorporation.

Although it is often desirable to study undisturbed natural assemblages of algae, it is also possible that some responses will be masked by the thick mats which rapidly attenuate light intensity. It is difficult, for example, to interpret P versus I curves conducted on the intact assemblage because the algae in the surface layer may be suffering photoinhibition while at the same time algae deeper in the mat may not have yet reached their optimum light intensity for photosynthesis. This problem was avoided in the above study by homogenizing the mat into a cell suspension. In an earlier study with an intact mat, however, significant photoinhibition was not observed.

There may also be ecotypic differentiation within an algal mat. The sun-shade characteristics of *Plectonema notatum* isolates from the upper and lower portions of a 1 cm-thick mat were compared (Sheridan 1979). Although each isolate exhibited a capacity for physiological adaptation to light intensity, the isolates from the bottom of the mat exhibited stable (presumably genetic) characteristics of shade adaptation when compared with isolates recovered from the top of the mat in July. Isolates recovered from the top of the mat during the winter, however, exhibited shade characteristics. Although the various isolates represent a continuum of ecotypes rather than distinct groupings, the results do illustrate a case in which the capacity for physiological adaptation within a species is superimposed on genetic variability.

Mastigocladus laminosus is perhaps the most cosmopolitan of thermophilic blue-green algae. It is heterocystous, exhibits true branching and occupies the upper thermal zone (up to 63–64°C) in hot springs which lack the high-temperature strains of *S. lividus*. In contrast to other thermophilic blue-greens, vegetative cells readily survive prolonged freezing or desiccation, an ability which no doubt accounts for the worldwide distribution of this alga. It colonized steam vents on the new volcanic island of Surtsey near Iceland within eight years of its formation. This species as well as several species of *Calothrix* are capable of active N_2^- fixation in hot springs. In one recent study (Wickstrom 1980) N_2 fixation in a thermal stream was limited to locations where these heterocystous algae were growing. That N_2 fixation activity was well adapted to prevailing temperatures was demonstrated by measuring the temperature response of N_2 fixation in algal mats which had been growing at 30°C and 40°C (Fig. 8.4). N_2 fixation in cell suspensions prepared from mats increased with increasing light intensity up to full sunlight.

At lower temperatures in the thermal gradient (below 55–60°C) there is a rapid increase in the species diversity of blue-green algae as well as the appearance of eukaryotic algae such as diatoms and *Chlorella* near 40°C.

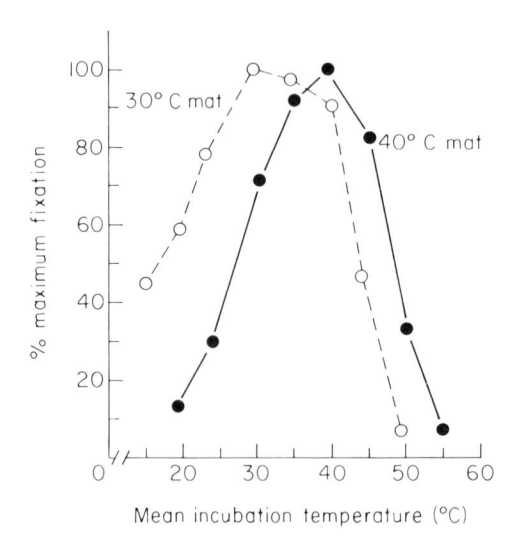

Fig. 8.4 Temperature response of nitrogen fixation in natural *Calothrix* mats which had been growing at 30°C and 40°C. (After Wickstrom 1980.)

Interspecific interactions begin to influence algal distribution in these more complex communities. Highly motile trichomes of *Oscillatoria terebriformis* are able to override and presumably shade out *Synechococcus* at temperatures (53–54°C) where strains of the latter alga grow well. The grazing activity of a thermophilic ostracod is thought to be responsible for the coexistence of *Pleurocapsa* and *Calothrix* in a distinctive nodular mat in Oregon hot springs. *Calothrix*, with a growth rate approximately twice that of *Pleurocapsa* and a vertically oriented growth form, would be expected to outcompete the slower-growing, prostrate *Pleurocapsa*. The ostracod, however, has a pronounced feeding preference for *Calothrix* and as a result, in hot springs populated by this particular ostracod, *Calothrix* is only found in association with *Pleurocapsa*. Adult and larval ephyrid flies are active grazers on the algal mats below 50°C in many hot springs in Yellowstone National Park (Wyoming, USA). Although the algal standing crop (measured as chlorophyll *a*) in this grazed area is lower, the photosynthetic efficiency is higher than further upstream where grazers do not occur. It is not clear whether the increased efficiency is due to increased nutrient cycling or to the dissected nature of the mat which would facilitate diffusion of gases and penetration of light.

Acid hot springs, habitats which cannot support the growth of blue-green algae, are colonized by an unusual eukaryotic alga, *Cyanidium caldarium*. Its taxonomic affinities are obscure, but recent opinion places it in the Rhodophyceae. This non-motile, coccoid organism has a *Chlorella* type of cell cycle, contains chlorophyll *a* and phycocyanin, has a broad pH optimum between 2 and 4 and does not grow at pH 5 or above. *Cyanidium* has a temperature optimum for growth and photosynthesis at 45°C and a range from near 20°C to 55–56°C. In natural habitats it replaces other acidophilic eukaryotic algae above 35–40°C. Temperature adaptation, such as is found in the thermophilic blue-green algae, is not exhibited by *Cyanidium* nor have different temperature strains been found. The organism is capable of heterotrophic growth on a large number of carbon sources although the ecological significance of this capacity has not been demonstrated.

REFERENCES

BOLD H.C. (1970) Some aspects of the taxonomy of soil algae. *Ann. N.Y. Acad. Sci.*, **175**, 601–16.

BRISTOL ROACH B.M. (1927) On the algae of some normal English soils. *J. agr. Sci.*, **17**, 563–88.

BROADY P.A. (1979) Qualitative and quantitative observations on green and yellow-green algae in some English soils. *Br. phycol. J.*, **14**, 151–60.

BROCK T.D. (1978) *Thermophilic Micro-organisms and Life at High Temperatures*. Springer-Verlag, New York.

CASTENHOLZ R.W. (1977) The effect of sulfide on the blue-green algae of hot springs. II. Yellowstone National Park. *Microbial Ecol.*, **3**, 79–105.

HOHAM R.W. (1980) Unicellular chlorophytes—snow algae. In Cox E.R. (ed.), *Phytoflagellates*, pp. 61–84. Elsevier/North-Holland, New York.

HOSTETTER H.P. & HOSHAW R.W. (1970) Environmental factors affecting resistance to desiccation in the diatom *Stauroneis anceps*. *Am. J. Bot.*, **57**, 512–18.

HUNT M.E., FLOYD G.L. & STOUT B.B. (1979) Soil algae in field and forest environments. *Ecology*, **60**, 362-75.
MADIGAN M.T. & BROCK T.D. (1977) Adaptation by hot spring phototrophs to reduced light intensities. *Arch. Microbiol.*, **113**, 111-20.
MEEKS J.C. & CASTENHOLZ R.W. (1971) Growth and photosynthesis in an extreme thermophile, *Synechococcus lividus* (Cyanophyta). *Arch. Mikrobiol.*, **78**, 25-41.
METTING B. (1981) The systematics and ecology of soil algae. *Bot. Rev.*, **47**, 195-312.
SHERIDAN R.P. (1979) Seasonal variation in sun-shade ecotypes of *Plectonema notatum* (Cyanophyta). *J. Phycol.*, **15**, 223-6.
STARKS T. L., SHUBERT L.E. & TRAINOR F.R. (1981) Ecology of soil algae: A review. *Phycologia*, **20**, 65-80.
WICKSTROM C.E. (1980) Distribution and physiological determinants of blue-green algal nitrogen fixation along a thermogradient. *J. Phycol.*, **16**, 436-43.

9 Algae and mankind

Algae have been the object of little applied research because they do not cause as many problems for mankind as do bacteria and fungi. This situation is likely to change in the future as humans and algae interact more often in both undesirable and desirable ways. Wastes from our cities, agriculture and industry will continue to pollute rivers, lakes, reservoirs and coastal waters as we crowd more and more people onto this planet. The effects on aquatic organisms are many, including an alteration of the competitive interactions in algal communities that often result in blooms of nuisance or toxic algae. At the same time we are learning how to take advantage of algal growth in controlled eutrophic systems to benefit mankind. Aquaculture of algae for food, nitrogen fixation, biomass conversion and sewage treatment are schemes presently under investigation. Basic research on the physiological ecology of algae is both benefiting and being benefited by the progress in these more applied areas.

TROUBLESOME ALGAE

The production of toxins by bloom-forming blue-green algae and red-tide-forming dinoflagellates has been mentioned in Chapter 4; a recent review (Collins 1978) considers a variety of algal toxins affecting multicellular organisms. The best known of the dinoflagellate toxins, saxitoxin, is a very potent inhibitor of the nervous system whose production appears to be limited to certain species in the genus *Gonyaulax* (Schmidt & Loeblich 1979). Toxin production is stimulated several fold during growth at higher salinities. Paralytic shellfish poisoning is not limited to waters with red tides because shellfish can accumulate toxic levels of the poison at *Gonyaulax* concentrations of about 200 cells ml^{-1} which do not colour the water red. The first symptom of poisoning is numbness about the mouth and fingertips; if enough toxin is ingested, death results from respiratory paralysis within 2–12 h. There is now evidence that the toxin also may be accumulated by herbivorous zooplankton and thereby cause fish kills (White 1981). The better known agent of fish kills is the red-tide dinoflagellate of the west coast of Florida (USA), *Gymnodinium breve*. The toxin involved, which is not as well characterized as saxitoxin, also causes neurotoxic symptoms (but no reported deaths) in humans who eat contaminated shellfish and is responsible for skin and eye irritation in swimmers.

Toxic characteristics are exhibited by certain species of at least eight genera of blue-green algae, but three species (*Microcystis aeruginosa*, *Anabaena flos-aquae*, *Aphanizomenon flos-aquae*) are the most serious offenders (Collins 1978). Waterfowl, fish and livestock are the most frequently affected organisms although human deaths have been reported. The toxins produced by each of these three species are different; *Aphanizomenon flos-aquae* produces saxitoxin in addition to three other uncharacterized toxins. In all three species toxin production is affected by environmental parameters. Bacteria isolated from toxic blooms of *Anabaena flos-aquae* depressed toxin production by this alga three-fold without affecting algal production (Carmichael & Gorham 1977). Another factor affecting bloom toxicity for all three species is the relative proportion of toxic and nontoxic strains in the population. In *Anabaena flos-aquae* both toxic and nontoxic clones can be isolated from individual colonies and are stable in culture. In cultures established from whole colonies containing both toxic and nontoxic strains, the nontoxic strains tend to dominate the culture after several years of subculture.

There are a number of other examples of toxic algae that cause problems but are not necessarily life-threatening to humans. The prymnesiophyte, *Prymnesium parvum*, is a small, marine flagellate that produces several toxins which are selectively toxic to gill-breathing animals. This organism is a serious pest for fish culture in Israel. *Prototheca*, a colourless relative of *Chlorella*, may cause infections in humans. At least two marine species of *Lyngbya*, a filamentous blue-green, produce toxins that cause dermatitis in humans.

Fouling, the settling and growth of attached organisms on submerged surfaces, is a serious economic problem. Shipping costs in time lost or in fuel consumption are increased significantly by the increased frictional resistance of even a small amount of fouling on a ship's hull. Fouling on offshore oil and gas platforms increases the loading forces caused by waves and currents and could result in structure failure if the design specifications are exceeded. Both algae and invertebrates are involved in fouling, but algae are the principal problem on modern supertankers because they can survive the rapid temperature changes experienced as these ships move back and forth between tropical and temperate waters (Evans 1981). *Enteromorpha*, *Ulothrix*, *Ectocarpus* and diatoms are some of the more troublesome forms. Copper-based antifouling paint has been used for many years as a means of reducing fouling. Copper leaches from the paint at a controlled rate of approximately 10 μg Cu^{++} cm^{-2} d^{-1}, maintaining toxic concentrations in the boundary layer next to the painted surface. Modern antifouling paints include organo-tin compounds such as triphenyltin chloride to increase the effectiveness of copper-based paints (Evans 1981).

In spite of progress in the development of antifouling compositions, fouling remains a problem because some algae are more resistant than others. The resistance of *Achnanthes* (a pennate diatom) to antifouling compounds is due in part to the production of a mucilaginous stalk that serves to raise the cell above the boundary layer containing toxins. In other diatoms mucilage capsules around the cells may afford some protection by forming copper complexes. Ultrastructural studies of *Navicula* and *Amphora* cells collected from test panels of copper-based paint revealed electron-dense bodies in the vacuole and cytoplasm that contained copper along with phosphorus, calcium and sulphur (Daniel & Chamberlain 1981). These bodies are thought to be important in maintaining low intracellular concentrations of free copper and therefore to help explain the tolerance of these diatoms to copper.

In the filamentous brown alga, *Ectocarpus siliculosus*, both copper-tolerant and copper-sensitive strains have been isolated (Hall 1981). Although both strains accumulate copper against a concentration gradient, the sensitive strains contain between one and seven times more copper than the tolerant strains at the same external levels of copper. Copper exclusion by the tolerant strains is due to properties of the cell membrane rather than to extracellular binding of copper. There is also evidence for intracellular immobilization of copper in the tolerant strains, although it has not been determined whether copper-containing bodies such as those described in diatoms are involved.

There is frequently a need to control algal growth which is stimulated in freshwater systems subjected to increased nutrient input from agriculture or domestic sewage. In addition to the health hazards of toxic blue-green algal blooms, nuisance growths of algae may spoil recreational uses of lakes and streams (e.g. *Cladophora*), clog filters in water treatment plants (e.g. diatoms) or impart disagreeable odours and tastes to municipal water supplies (e.g. *Synura*). Of course the best method of control is to eliminate the source of the extra nutrients, but this is a long-term solution that is usually expensive and often involves political considerations. The case of Lake Washington (USA) is proof that major pollution problems can be solved if a concerted effort is made (see Edmondson & Lehman 1981). Short-term, politically expedient solutions are more common. The usual approach is to use an algicide. Copper sulphate, applied at a rate of 0.2–2 mg l^{-1}, is the most commonly used algicide although a number of other compounds are also in use (Trainor 1978).

EFFECTS OF POLLUTANTS ON ALGAE

In many of the cases discussed above the troublesome algae are the dominant species in an 'unnatural' algal community of low-species diversity resulting from some major human-related perturbation. In this section, the effects of pollutants on unperturbed algal communities are discussed. For example, what are the effects of chronic, low-level exposure to heavy metals, pesticides and herbicides? At the other end of the spectrum, what is the short-term impact and rate of recovery from an isolated event such as a major oil spill?

After the emotional impact of dying, oil-soaked birds, it is somewhat surprising to learn that algae are *relatively* unaffected by major oil spills (see review by O'Brien & Dixon 1976). The most toxic components in crude oil are the volatile aromatic compounds such as benzene, toluene and xylenes. If these compounds have time to evaporate before the spill reaches shore, damage to subtidal and intertidal seaweeds is much less severe. The physiological effects of oil include the disruption of cellular membranes, as yet uncharacterized effects on other intracellular components and the restriction of the diffusion of inorganic carbon into plants covered by a film of oil. Species vary in their sensitivity. Thick deposits of oil stranded in the upper intertidal zone may smother algae and restrict recolonization of the area. Intertidal invertebrates are generally more sensitive than algae and as a result certain algae have increased in abundance following a spill because of the absence of grazers. In the case of the massive *Torrey Canyon* spill off the Cornish coast of England in 1967, the dispersants used to clean up the mess did most of the damage (Southward & Southward 1978). Although algae such as *Enteromorpha*, *Ulva* and *Fucus* returned to the heavily damaged sites within six months, it has taken close to ten years for the normal, pre-spill community to develop.

The effects of low-level, sublethal pollution are much less obvious than those from oil spills but are probably of greater ecological significance in the long term. The 'big-bag' systems (p. 161) for enclosing natural planktonic communities are especially useful for ecologically meaningful studies of this problem. The Controlled Ecosystem Pollution Experiment (CEPEX) is a good example of a large, multidisciplinary study of this type (*Bulletin of Marine Science* 1977). The major impact of additions of either copper or petroleum products was a replacement of centric diatoms with pennate diatoms and microflagellates. Associated with this change in community structure was a decline in phytoplankton biomass and productivity for several days following the additions and then a gradual increase, so that by the end of the four-week experiment these parameters were similar in treated and control enclosures.

There are many examples of laboratory studies on the effects of various heavy metals and organic pollutants on phytoplankton. In most of these studies a single, high-level dose of the pollutant is added and the organism's response is measured within a few hours or days. While this approach is invaluable as a means of screening pollutants and comparing the sensitivity of different species, it does not mimic the field situation where pollution is often chronic and low level. The big-bag system is a useful approach, as indicated above, but is a major undertaking and provides little information on the physiological responses of individual species, especially the sensitive species which are replaced early in the experiment. Chemostat cultures are ideally suited for the study of these low-level, long-term effects of pollutants.

In one such study (Cloutier-Mantha & Harrison 1980) the addition of sublethal concentrations of mercury every 8–12 h to nitrogen-limited chemostat cultures of *Skeletonema costatum* did not affect the growth rate for several days, after which the growth rate declined at 0.37 nmol l^{-1} Hg and dropped to zero at 3.7 nmol l^{-1} Hg. After 16 days of this semi-continuous exposure, however, there was a rapid recovery at both Hg concentrations and after three weeks, cell densities in the control and both of the treated chemostats were similar. The half-saturation constant, K_s, for NH_4^+ uptake showed a similar decline and recovery to control values, whereas the maximum rate of NH_4^+ uptake, V_m, was relatively unaffected. These results suggest that during the initial (one to two weeks) exposure to Hg, *Skeletonema costatum* has a greatly reduced growth capacity at low nitrogen concentrations. There is evidence that heavy metal toxicity is caused by metal binding to sulphydryl-containing compounds in the cell (Fisher & Jones 1981).

In the course of studies on the effects of pollutants on phytoplankton, it has been observed that isolates from oceanic environments tend to be more sensitive to pollutants than isolates of the same species obtained from coastal waters. In recent studies (Murphy & Belastock 1980) sensitivity to organic pollutants differed among conspecific coastal isolates and was inversely correlated with the amount of pollution in the habitat from which they had been isolated. Although these differences presumably reflect genetic adaptations, the chemostat study discussed above shows that rapid adaptation to pollutants does occur. The possibility of physiological adaptation has not been ruled out.

MASS CULTIVATION OF ALGAE

The Japanese have included seaweeds, especially *Porphyra*, in their diets for several hundred years. They often placed brush in the intertidal zone to provide additional substrate for the alga. Following the discovery by Kathleen Drew that the filamentous, shell-inhabiting alga, *Conchocelis*, is an alternate phase in the *Porphyra* life history, the '*Conchocelis*' phase has been cultured on shells in the *Porphyra* beds. This procedure results in increased spore settlement on the artificial substrates (nets suspended from bamboo poles are frequently used nowadays) and a more predictable harvest. More recently there has been increased interest in other countries in the mass cultivation of algae, both microalgae and seaweeds. These research programmes are directed toward various goals including: production of food for human consumption; feed production for livestock and for aquaculture systems; hydrocolloid production; production of bulk, algal, organic matter for conversion to methane; and the use of algal mass cultures as nutrient sinks in tertiary, sewage-treatment systems.

As is the case in the more usual forms of agriculture the maintenance of a highly productive monoculture of a desirable alga is associated with a variety of costs in terms of energy, labour, and resources. Simply encouraging the growth of a particular alga in its natural habitat, as the Japanese do with *Porphyra*, is clearly the least-expensive system in terms of materials and energy but is highly labour intensive. Although production is lower and competition and predation take their toll, nature is meeting the alga's needs for free. Higher production requires the more controlled, and therefore expensive, conditions of a land-based container of some type. In practice, large-scale cultures are maintained outdoors to take advantage of the available sunlight, nutrients are provided in excess and the culture is mixed. Under these conditions, temperature and sunlight (intensity and duration) are the only two uncontrollable environmental parameters. In temperate regions, temperature control during the winter months could be achieved by utilizing the thermal effluent from a power-generating plant. In recently studied applications, overall costs are reduced in integrated multi-use systems in which the algal biomass produced during tertiary waste-water treatment is used for feed or methane production. In industrialized countries, waste-water often contains toxic components which must not be allowed to enter important food chains.

It should be noted that the usual objective in algal mass culture is to obtain maximum yields ($g\ m^{-2}\ d^{-1}$) and not necessarily maximum specific growth rates ($g\ g^{-1}\ d^{-1}$). Since the highest specific growth rates are usually obtained at low population densities and since yield is the product of specific growth rate and density, it is obvious that the highest yields will be obtained at intermediate densities and growth rates.

Microalgae

The applications and theoretical aspects of the mass culture of microalgae have been considered in several recent reviews (Goldman 1979a, 1979b; Soeder 1980). Large-scale microalgal cultures are operated as continuous or semi-continuous cultures in which nutrient solution is introduced at one end of the system and the algae are harvested as the culture flows out the other end. Mixing is provided by aeration, baffles, paddles or meandering channels. Losses to planktonic grazers are not a major problem because the brief residence time required for growth of the single-celled algae does not allow slower-growing consumers to proliferate. Controlling the species composition of the culture continues to be a problem. The size and openness of these systems precludes axenic conditions and weed algae such as *Chlorella* and *Scenedesmus* in freshwater systems and *Phaeodactylum* and *Skeletonema* in sea-water cultures tend to become dominant. Species control is less of a problem in the mass culture of *Spirulina* (for human consumption) and *Dunaliella* (for glycerol production) due to the extreme chemical environments in which these algae thrive. Harvesting microalgae is a costly operation and is usually accomplished by filtration (especially for *Spirulina* which is filamentous) or by flocculation and flotation rather than by centrifugation. Drying and processing follow.

Despite the enthusiasm in the years following World War Two for microalgae as a source of protein for a hungry world, it now appears unlikely that microalgae will make a significant contribution to this serious problem. The costs are too high and the technology too advanced for extensive use in underdeveloped countries where the need is greatest. In some countries such as Japan, however, there is a large market for high-priced health foods and culturing microalgae for human consumption can be a profitable venture. Even if microalgal protein could be produced simply and economically, there remains the problem of having a new food item, such as *Chlorella* tablets, accepted into the diet. The Japanese are an exception because of their long tradition of eating algae. A more likely use of microalgae is as part of a multipurpose system such as a sewage treatment plant in which an algal–bacterial biomass produced on waste-water is utilized by filter-feeding fish or bivalves which are suitable for human consumption. The algae utilize CO_2 and nutrients which the bacteria release as they mineralize the organic matter. The O_2 released by the algae enhances the bacterial decomposition process. This mutualistic association largely eliminates the need for aeration. The use of filter feeders to harvest the algal–bacterial biomass eliminates the costs of harvesting, drying and processing the microalgae.

Seaweeds

Because of their larger size, seaweeds are readily harvested by hand and have a long history of being used for human consumption, livestock feed, as manure in agriculture and as sources of soda, potash and iodine (Chapman 1970). More recently the commercial applications of seaweeds are largely for human consumption and for the production of the hydrocolloids agar, carrageenan and algin which are widely used for their emulsifying and gelling properties. In cases such as the harvest of natural *Macrocystis* beds along the coast of California (USA), specially designed ships efficiently cut the plants a few metres below the surface and recover the cut portions on a conveyor belt. Because of their attached-growth habit seaweeds may be cultivated in their natural habitat on recoverable, artificial substrates such as ropes or nets. Larger species may be tied to the substrate. If the substrates are suspended some distance above the sediment, the crop will be less accessible to sea urchins and other bottom-dwelling grazers and cultures need not be restricted to areas of shallow water. In some species the holdfast remains attached to the substrate and will produce a new plant in the following year, eliminating the need to reinoculate the substrate.

Seaweeds may also be cultivated for their hydrocolloids in land-based culture systems, although most such projects are still in the research stage (Jensen & Stein 1979; Krauss 1977). Material and energy costs are higher than for field cultivation, but these systems are less labour intensive and may be incorporated into sewage treatment systems to further reduce overall costs. Extensive growth of diatoms and other epiphytes, which is often a problem in seaweed culture, may be controlled by increasing the turbulence and rate of water exchange in the system or by mild chlorine treatment. Without their natural predators, grazers may become serious pests in culture. Control in *Chondrus crispus* cultures has been achieved by introducing fish predators of the crustacean grazers and by passing the plants between rubber rollers to crush snail grazers without damaging the algae (Shacklock & Croft 1981). In spite of these problems, seaweed cultures offer several advantages. In the case of *Gigartina* and *Chondrus* it has been possible to select fast-growing strains that can be propagated vegetatively. It seems likely that further selection and breeding could improve the yield and chemical properties of the hydrocolloid produced. It is also possible to manipulate the environment in cultures to enhance hydrocolloid production; several species of red algae have an increased hydrocolloid content when grown for a period of time at low nitrogen concentrations.

REFERENCES

BULLETIN OF MARINE SCIENCE (1977) 27, 1–175. (Several papers on the Controlled Ecosystems Pollution Experiment.)
CARMICHAEL W.W. & GORHAM P.R. (1977) Factors influencing the toxicity and animal susceptibility of *Anabaena flos-aquae* (Cyanophyta) blooms. *J. Phycol.*, 13, 97–101.
CHAPMAN V.J. (1970) *Seaweeds and Their Uses*. Methuen & Co., London.
CLOUTIER-MANTHA L. & HARRISON P.J. (1980) Effects of sublethal concentrations of mercuric chloride on ammonium-limited *Skeletonema costatum*. *Mar. Biol.*, 56, 219–31.
COLLINS M. (1978) Algal toxins. *Microbiol. Rev*, 42, 725–46.
DANIEL G.F. & CHAMBERLAIN A.H.L. (1981) Copper immobilization in fouling diatoms. *Bot. Mar.*, 24, 229–43.
EDMONDSON W.T. & LEHMAN J.T. (1981) The effect of changes in the nutrient income on the condition of Lake Washington. *Limnol. Oceanogr.* 26, 1–29.
EVANS L.V. (1981) Marine algae and fouling: A review, with particular reference to ship fouling. *Bot. Mar.*, 24, 167–71.
FISHER N.S. & JONES G.J. (1981) Heavy metals and marine phytoplankton: Correlation of toxicity and sulfhydryl-binding. *J. Phycol.*, 17, 108–11.
GOLDMAN J.C. (1979a) Outdoor algal mass cultures. I. Applications. *Water Res.*, 13, 1–19.
GOLDMAN J.C. (1979b) Outdoor algal mass cultures. II. Photosynthetic yield limitations. *Water Res.*, 13, 119–36.
HALL A. (1981) Copper accumulation in copper-tolerant and non-tolerant populations of the marine fouling alga, *Ectocarpus siliculosus* (Dillw.) Lyngbye. *Bot. Mar.*, 24, 223–8.
JENSEN A. & STEIN J.R. (eds) (1979) *Proceedings of the Ninth International Seaweed Symposium*. Science Press, Princeton.
KRAUSS R.W. (ed.) (1977) *The Marine Plant Biomass of the Pacific Northwest Coast*. Oregon State University Press, Cornwallis, Oregon.
MURPHY L.S. & BELASTOCK R.A. (1980) The effect of environmental origin on the response of marine diatoms to chemical stress. *Limnol. Oceanogr.*, 25, 160–5.
O'BRIEN P.Y. & DIXON P.S. (1976) The effects of oils and oil components on algae: A review. *Brit. phycol. J.*, 11, 115–42.
SCHMIDT R.J. & LOEBLICH A.R. III (1979) Distribution of paralytic shellfish poison among Pyrrhophyta. *J. mar. Biol. Ass. U.K.*, 59, 479–87.
SHACKLOCK P.F. & CROFT G.B. (1981) Effect of grazers on *Chondrus crispus* in culture. *Aquaculture*, 22, 331–42.
SOEDER C.J. (1980) Massive cultivation of microalgae: Results and prospects. *Hydrobiol.*, 72, 197–209.
SOUTHWARD A.J. & SOUTHWARD E.C. (1978) Recolonization of rocky shores in Cornwall after use of toxic dispersants to clean up the *Torrey Canyon* spill. *J. fish. Res. Bd. Can.*, 35, 682–706.

TRAINOR F.R. (1978) *Introductory Phycology.* John Wiley & Sons, New York.
WHITE A.W. (1981) Marine zooplankton can accumulate and retain dinoflagellate toxins and cause fish kills. *Limnol. Oceanogr.*, **26**, 103–9.

Appendix 1
Detailed characteristics of algal classes

F = flagellation in the motile cell.
CW = the structural component of the cell wall.
FR = food reserve.
E = the eyespot (stigma) in the motile cell.
Chl = chlorophyll(s).
AP = the accessory pigments, xanthophyll or phycobilins;' only major, characteristic pigments are given.
T = the number of photosynthetic thylakoids associated in a band.
GL = the girdle lamellae, a band of thylakoids encircling the chloroplast, just inside the chloroplast envelope.
CER = the chloroplast endoplasmic reticulum, a layer of endoplasmic reticulum surrounding the chloroplast, resulting in four membranes around the chloroplast; in some groups the outer membrane is contiguous with the outer membrane of the nuclear envelope.

Cyanophyceae F—absent, some move by gliding; CW—complex mucopeptide similar to that in Gram-negative bacteria; FR—starch; E—absent; Chl—a only; AP—phycobilins, contained in phycobilisomes; T—single; GL—absent; CER—absent.

Chlorophyceae F—two, four, or more, equal and smooth with apical insertion: the motile cell is symmetrical; CW—absent, cellulose or other polysaccharide; FR—starch in the chloroplast: all other groups, except the Charophyceae, store the food reserve outside the chloroplast; E—in the chloroplast, not associated with the flagella; Chl—*a* and *b*; T—various; GL—absent; CER—absent.

Charophyceae F—two, equal and smooth with subapical insertion: the motile cell is asymmetrical; CW—absent, cellulose or other polysaccharide; FR—starch in the chloroplast: all other groups, except the Chlorophyceae, store the food reserve outside the chloroplast; E—absent; Chl—*a* and *b*; T—various; GL—absent; CER—absent.

Euglenophyceae F—basically two with anterior insertion, usually only one is emergent with a unilateral row of fine hairs; CW—pellicle of interlocking proteinaceous strips; FR—paramylon; E—in the cytoplasm, associated with the swelling on the emergent flagellum; Chl—*a* and *b*; T—in threes; GL—absent; CER—? only one extra membrane surrounds the chloroplast envelope.

Pyrrhophyceae F—two, with lateral insertion: one is circumferential and highly specialized with fine hairs, the second flagellum is trailing and smooth; CW—theca of flattened vesicles, filled to various extent with cellulose; FR—starch and lipid; E—various, some highly complex; Chl—*a* and *c*; AP—peridinin; T—in threes; GL—absent; CER—? only one extra membrane surrounds the chloroplast envelope.

Xanthophyceae F—two, unequal and with anterior to lateral insertion: the longer have bilateral rows of stiff hairs and are directed anteriorly: the shorter are smooth and are directed posteriorly; CW—cellulose; FR—lipid and chrysolaminaran; E—in the chloroplast and associated with swelling on the smooth flagellum; Chl—*a* and *c*; T—in threes; GL—present; CER—present and connected to the nuclear envelope.

Chrysophyceae F—two, unequal and with anterior to lateral insertion: the longer have bilateral rows of stiff hairs and are directed anteriorly, the shorter are smooth and are directed posteriorly; CW—absent: some produce loricas or siliceous scales: siliceous statospore is characteristic of group; FR—lipid and chrysolaminaran; E—in the chloroplast and associated with swelling on the smooth flagellum; Chl—*a* and *c*; AP—fucoxanthin; T—in threes; GL—present; CER—present and connected to the nuclear envelope.

Bacillariophyceae F—one, directed anteriorly subapical insertion and bilateral rows of stiff hairs: 9 + o arrangement of the microtubules: only the flagellated cell is the male gamete in centrics; CW—silica; FR—lipid and chrysolaminaran; E—absent; Chl—*a* and *c*; AP—fucoxanthin; T—in threes; GL—present; CER—present and connected to the nuclear envelope.

Phaeophyceae F—two, unequal and with lateral insertion: the longer have bilateral rows of stiff hairs and are directed anteriorly, the shorter are smooth and are directed posteriorly; CW—cellulose; FR—laminaran; E—in the chloroplast and associated with swelling on the smooth flagellum; Chl—*a* and *c*; AP—fucoxanthin; T—in threes; GL—present; CER—present and connected to the nuclear envelope.

Eustigmatophyceae F—one with anterior insertion and bilateral rows of stiff hairs: the second basal body is present; CW—cellulose?; FR—lipid?; E—large, in the cytoplasm and associated with the hairy flagellum; Chl—*a* only; T—in threes; GL—absent; CER—present but not connected to the nuclear envelope.

Prymnesiophyceae F—two, equal or subequal, smooth and usually directed posteriorly during swimming: unique haptonema attached between the flagella; CW—cellulosic scales, often calcified on the outer rim to form coccoliths; FR—lipid and chrysolaminaran; E—absent except in one genus and not associated with the flagellum; Chl—*a* and *c*; AP—fucoxanthin; T—in threes; GL—absent; CER—present and connected to the nuclear envelope.

Crytophyceae F—two and anterior to subapical insertion: one has a bilateral row of stiff hairs and the other has a unilateral array of shorter hairs; CW—thin, proteinaceous periplast; FR—starch and lipid; E—part of the chloroplast, near the nucleus and not associated with the flagella; Chl—*a* and *c*; AP—phycobilins, contained in the lumen of the thylakoids; T—in pairs; GL—absent; CER—present and connected to the nuclear envelope.

Rhodophyceae F—absent; CW—cellulose; FR—floridean starch; E—absent; Chl—*a* and *d*; AP—phycobilins in phycobilisomes; T—single; GL—present; CER—absent.

Appendix 2
Methods in algal physiological ecology

Simply stated, the basic techniques available to algal physiological ecologists, whether they are working with natural systems or cultures, are designed to answer one or more of the following four questions: Who's there? How many of them are there? What are they doing? How fast are they doing it? In more scientific language the first question concerns the identity (genus and species) of the organisms in question. The answers may be obtained with a good microscope, proper taxonomic keys and some experience and will not be considered at length here. The second question may also be addressed with the light microscope, but like the first, the effort can be tedious and time-consuming when dealing with natural populations of microscopic algae. Estimates of total algal biomass are often used to answer the second question. The third and fourth questions usually relate to photosynthesis and its manifestation in growth. The emphasis is not unexpected since primary production by algae is the base of the food chain in most aquatic systems. More recently other activities of algae such as nutrient uptake, nitrogen fixation, respiration and heterotrophy have been receiving more attention.

On the following pages, biomass, primary productivity and culture methods are discussed in general terms. The discussion takes the general view that one is dealing with phytoplankton or a suspension of cultured cells; variations as they apply to algae from different habitats are considered in the appropriate chapter. The reader is referred to the following references for details on these and other methods used in algal physiological ecology: Hellebust and Craigie (1978); Stein (1973); Strickland and Parsons (1972); and Vollenweider (1969).

BIOMASS (STANDING CROP)

It is often useful to express rate functions on either an area or volume basis or on some measure of biomass or standing crop. In culture work cell number is frequently used as a basis for reporting data because it is a meaningful parameter that is relatively easy to measure, especially with electronic particle counters. In natural systems, however, cell number is more difficult to measure because the cells are not as concentrated; electronic particle counters cannot be used as they do not distinguish living cells from dead cells and detritus particles. Cell-number data obtained through microscopic examinations are useful for documenting community structure and for temporal and spatial comparative studies, but are less useful as a measure of algal biomass. Due to the extreme variation in cell volume among algal species, it is difficult to convert cell number to biomass without knowing the average cell volume and vacuole size of each species. In the case of unicellular organisms reproducing by binary fission, any result expressed on a per cell basis could vary by a factor of two during a single cell cycle. In a randomly dividing population the effect will average out to some mid-point of the cell cycle, but in a population in which division is partially synchronized (as often happens in nature) this cell-cycle effect could be significant. In laboratory work on pure cultures, corrections for cell cycle effects can be obtained by monitoring the growth curve of the culture.

Biomass or standing crop is often expressed as wet weight, dry weight, ash-free dry weight, energy equivalents, total carbon, protein or chlorophyll. Any of these values may be readily obtained from a pure culture since the organism under study is the only particulate matter present. In natural systems, the algae are accompanied by other living organisms and by large quantities of non-living, particulate, organic and inorganic matter whose size range is fully as great as that represented by the algae present. Since no practical technique has been devised to physically separate the algae from other particulate matter, aquatic ecologists have searched for some cell component which could serve as a reliable index of algal biomass. Ideally this compound would be unique to the algae, easy to measure at low concentrations, present in measurable amounts, degraded rapidly in dead cells and present in constant ratios to other ecologically important cell parameters. To date, chlorophyll a comes closest to meeting these criteria. In addition it has the advantage of being intimately involved in photosynthesis, a very important ecological function performed by algae. The principal difficulties encountered with chlorophyll are: (1) degradation products are included in most chlorophyll measurements giving falsely high values; (2) ratios of chlorophyll to other cell components are far from constant; and (3) the total algal assemblage, rather than the individual species in question, is measured.

For chlorophyll measurements, the particulate matter is concentrated, usually by filtration, and extracted in 90% acetone. The absorbance of the extract is measured at the red peak of chlorophyll (665 nm) in a spectrophotometre and converted to μg chlorophyll a per volume of sample using the appropriate extinction coefficient and conversion factors. Chlorophylls b and c have low absorbances at 665 nm and thus do not interfere significantly, although corrections for their presence can be made by taking readings at 645 nm and 630 nm and applying appropriate equations (Strickland & Parsons 1972). In these procedures all three of the chlorophyll a degradation products shown in Fig. A2.1 will be included in the chlorophyll a measure-

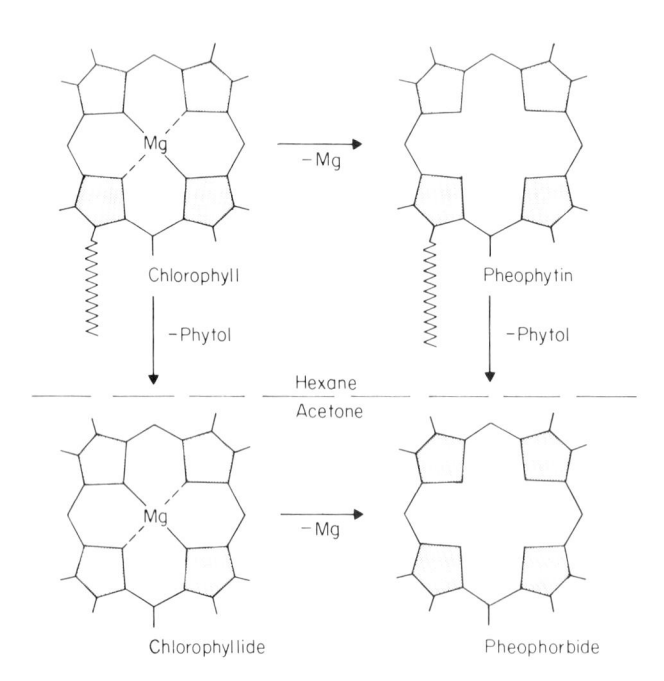

Fig. A2.1 Chlorophyll and its degradation products.

ment. An acidification step is commonly employed to correct for the presence of pheophytin and pheophorbide. When acid is added to the extract, any chlorophyll a or chlorophyllide a present is converted to pheophytin a or pheophorbide a respectively, with the loss of Mg from the molecule. The conversion results in a shift in the absorption spectrum which is seen as a decrease in the absorbance at 665 nm. Since the decrease in absorbance is proportional to the amount of chlorophyll a and chlorophyllide a present, the combined amount of these compounds is readily calculated by measuring the absorbance of the extract before and after adding acid.

This technique has recently been refined (Whitney & Darley 1979) by partitioning the acetone extract with hexane prior to absorbance determinations. The presence of the phytol chain on chlorophyll *a* and pheophytin *a* renders the molecules very hydrophobic; as a result, they migrate to the hexane phase leaving the more hydrophilic chlorophyllide *a* and pheophorbide *a* in the acetone phase. Applying the acidification technique to the hexane phase yields a more accurate measure of the chlorophyll *a* actually present in the sample. The need for the application of the additional corrective procedures will depend on the extent of contamination by degradation products in the initial sample. Thin-layer chromatography (Jeffrey 1981) may be used for a more detailed analysis of pigments, including carotenoids, in phytoplankton samples.

Chlorophyll may also be measured by a fluorometric method which takes advantage of the fact that chlorophyll emits red light following the absorption of blue light. Since the intensity of the fluorescence is measured relative to darkness instead of against a light beam as in the spectrophotometric method, the fluorometric method is at least ten times more sensitive. Even though it is more difficult to correct for contaminating chlorophyll degradation products with this method, it would probably be the method of choice in oligotrophic waters with very low concentrations of phytoplankton.

Adenosine triphosphate (ATP) also meets many of the criteria mentioned above as an index of algal biomass. The principal problems are: (1) it is not unique to algae; and (2) the variation in ATP per cell can change dramatically under non-growing conditions. ATP measurements could prove useful, however, in systems dominated by healthy algae.

PRIMARY PRODUCTIVITY

Photosynthesis converts light energy into chemical energy which is stored in the form of organic matter (reduced carbon) in compounds such as carbohydrates, proteins and lipids. The amount of new organic matter produced by autotrophs is called primary production, whereas the amount produced per unit time (on a volume or area basis) is called primary productivity. Not all of the primary production is retained by the cell, however, because the plant or alga must respire some of the organic matter to provide energy to meet many of its metabolic needs. Thus, the total amount of carbon fixed is referred to as gross photosynthesis (or gross primary production) and the amount of carbon which is left over and stored by the cell is called net photosynthesis (or net primary production). In higher plants, net production is often about 50% of gross production due to the high maintenance cost of supporting structures such as roots and stems. In simpler organisms which usually lack these 'parasitic', non-photosynthetic tissues, net production is closer to 90% of gross production. Production may be expressed as energy equivalents by measuring calories of heat energy released upon combustion.

The most direct, and therefore the best, methods for measuring primary productivity are based on the summary equation for the photosynthetic reactions:

$$6CO_2 + 6H_2O \longrightarrow C_6H_{12}O_6 + 6O_2 \qquad \text{(Eqn A2.1)}$$

Perhaps the most obvious approach is to measure the increase in biomass ($C_6H_{12}O_6$) over time. This procedure is called the harvest method by terrestrial ecologists. Since respired carbon is not included, this method measures net productivity. The time interval involved is usually days or weeks since small changes in biomass are not easy to detect. A serious drawback of the method when used in natural systems is that any losses (e.g. due to grazing) which occur between the sample periods will result in an underestimation of primary productivity. With macroscopic algae, herbivorous activity may be minimized or eliminated with cages, but in phytoplankton work it is very difficult to correct for losses due to grazing, sinking and current transport. In addition, the difficulties in separating algae from other particulate matter (discussed in the previous section) are encountered here.

Since measuring the decrease in water content of the system is obviously impractical, we are left with the rates of CO_2 consumption and O_2 evolution as measures of primary productivity. Respiration by heterotrophs as well as by the algae themselves will evolve CO_2 and consume O_2 and therefore decrease the apparent productivity measured by these techniques. Changes in total CO_2 are not often used for productivity measurements in aquatic systems due to the high background of dissolved inorganic carbon (H_2CO_3, HCO_3^-, $CO_3^=$) in the water. Tagging the CO_2 consumed with the radioisotope ^{14}C, however, is a very sensitive method. Change in pH, which is a function of dissolved inorganic carbon, has also been used as a convenient, non-destructive way of monitoring photosynthesis and respiration in laboratory microcosms (Odum 1971).

O_2 production has been used extensively since 1927 as a measure of primary production in the 'light- and dark-bottle method'. Small changes in dissolved oxygen are readily measured titrametrically by the sensitive and reliable Winkler method, although the use of oxygen electrodes is increasing. To measure primary productivity in a water sample, replicate 300 ml subsamples are placed in three glass bottles (more bottles are used if duplicate measurements are to be made). Winkler reagents are added to one bottle for later determination of the initial oxygen concentration in the water sample. One of the remaining bottles is darkened and then it and the third, uncovered, light bottle are incubated in situ or in an incubator for 3 to 24 hours, after which metabolism is stopped by the addition of the Winkler reagents and the O_2 level is determined. The decline in O_2 level during the incubation period (that of the initial bottle minus that of the dark bottle) is taken as a measure of respiration by the total community in the enclosed volume of water over the time interval. Similarly, the increase in O_2 (that of the light bottle minus that of the initial bottle) is taken as a measure of net photosynthesis. The total change in O_2 level (that of the light bottle minus that of the dark bottle) is used as a measure of gross photosynthesis. Since productivity measurements are usually expressed in terms of carbon rather than oxygen, appropriate conversion factors have to be employed. The photosynthetic quotient (vol. O_2 liberated per unit vol. CO_2 consumed) is near 1.0 when carbohydrate is the product (see Eqn A2.1), but can be higher (e.g. 1.4) if more highly reduced products such as lipid are produced. An average photosynthetic quotient of 1.2 is frequently used with a respiration quotient (vol. CO_2 liberated per unit vol. O_2 consumed) of 1.0.

There are a number of limitations in this method which should be considered in addition to the problems of choosing a suitable photosynthetic quotient for the system under study. The method has proved useful in waters with reasonably high productivity (3–200 mg C m^{-3} h^{-1}) but is not sensitive enough for oligotrophic waters in which the long incubation periods required to obtain significant changes in O_2 levels also contribute to some of the artifacts mentioned below. In highly productive areas the water may become supersaturated with O_2 and oxygen could be lost when reagents are added at the end of the incubation period. The assumption is made in using this method that the respiratory rate in both the heterotrophs and autotrophs is not affected by light intensity and is therefore the same in the light and dark bottles. The validity of this assumption is still being debated; the error introduced, if any, in photosynthetic rates is probably slight. Enclosing a water sample, especially for periods of a day or more, may cause serious problems. Bacteria quickly colonize the glass surface and multiply rapidly at the expense of organic matter which tends to accumulate on surfaces. The increased bacterial respiration could result in an underestimate of net photosynthesis and alter a number of environmental parameters including nutrient concentrations. Decreased turbulence resulting from containment could affect the physiology of phytoplankton cells as they become concentrated at the bottom of the bottle. The species composition of the sample may also change within 24 h due to differences in the ability of various species to adapt to the modified environment within the bottle.

Measuring primary production by the incorporation of ^{14}C-labeled bicarbonate would appear to be the most direct method available because algal photosynthesis is the principal metabolic process involved. The technique is approximately 60 times more sensitive than the O_2 method and is therefore the only practical one to use in oligotrophic waters. Radioactive bicarbonate is added to water samples in light and dark bottles which are incubated for 1–6 h in situ or in an incubator, after which the cells (and other particulate matter) are recovered by filtration. The radioactivity contained on the filters is counted with a Geiger counter or by the more recently introduced liquid scintillation counting techniques. In order to convert the counts on the filter to mg C assimilated, one must know the fraction of the total inorganic carbon in the water sample which is radioactive (i.e. the specific activity or the total number of counts added and the total amount of inorganic carbon present). The total inorganic carbon may be determined from pH and alkalinity measurements or more directly by infrared spectroscopy of CO_2 liberated following acidification of the water sample. One must also know the counting efficiencies of the system one is using in order to convert counts per minute (as measured with the counting device) to disintegrations per minute (actual nuclear decay events).

$$\text{Carbon assimilated} = \frac{\text{d/min on filter} \times \text{total inorganic carbon}}{\text{total d/min}} \quad \text{(Eqn A2.2)}$$

The appropriate conversion factors and units are used to express carbon assimilation as mg C m^{-3} h^{-1}. A small correction (1.05) is often added to correct for metabolic discrimination against ^{14}C.

The containment phenomena mentioned above are also a problem with the ^{14}C method but are less severe due to the shorter incubation periods. The filtration process can rupture delicate cells, whose soluble contents are then washed through into the filtrate and lost from the counting process. Preserving samples with formaldehyde or other fixatives at the end of the incubation period prior to filtration can result in leakage of soluble cell components. The deposition of CaCO$_3$, such as occurs in coccolithophorids, will result in overestimates of production. The problem can be corrected before counting by exposing the filters to HCl fumes which will drive off the particulate inorganic carbon as CO$_2$. The filters themselves may retain unincorporated ^{14}C and must be decontaminated with HCl fumes or by some other process (Lean & Burnison 1979). Another problem encountered in the ^{14}C method relates to the release of dissolved organic carbon (DOC) by the cells, a phenomenon discussed in more detail in Chapter 3. The lost material may be counted in the filtrate by liquid scintillation counting, after acidifying and bubbling it to drive off unincorporated inorganic carbon as CO$_2$. Recently some investigators have modified the ^{14}C technique by acidifying and bubbling the entire water sample and then counting the DOC activity right along with the particulate activity in the cells (Theodórsson & Bjarnason 1975). A number of controls must be included in such measurements, especially those to correct for contaminating organic or particulate radioactivity in the radiocarbon stock solution.

Photosynthesis measured by the ^{14}C method is neither net nor gross photosynthesis but somewhere in between, probably closer to net photosynthesis. The uncertainty develops for two reasons. If cells are growing slowly it may take a relatively long time for the cells to reach isotopic equilibrium (i.e. when the specific activity of carbon in the cells equals the specific activity of carbon in the medium). As a result, the carbon respired early in the incubation period is enriched in ^{12}C relative to the carbon being fixed, thereby contributing to an overestimation of net productivity. Once isotopic equilibrium is reached, the specific activity of the fixed carbon will equal that of the respired carbon and net productivity will be measured. Secondly, some of the respired carbon is refixed by photosynthesis, causing various effects on the measurement, depending on the amount of carbon recycled and its specific activity. In spite of the many technical considerations and uncertainties involved, the ^{14}C technique is an invaluable tool for aquatic ecologists. One must simply keep its limitations in mind when interpreting data.

CULTURE TECHNIQUES

Batch culture is the simplest and most commonly employed culture system. Algal cells are inoculated into a nutrient solution contained in a suitable vessel which is often shaken or aerated to suspend the cells, replenish CO$_2$ and otherwise maintain a homogeneous environment. The growth curve (cell concentration versus time) exhibits the phases (lag, exponential, stationary, death) characteristic of such systems (Fig. A2.2). Due to the nature of exponential growth, the absolute cell concentration and the culture environment change relatively slowly during early exponential growth but very rapidly during late exponential growth. The familiar straight-line, exponential phase of the semi-log plot in Fig. A2.2 makes it easy to overlook this basic fact of batch cultures. A simple example of a potentially nitrate-limited culture, plotted on a linear scale, provides an instructive reminder (Fig. A2.3). The first doubling in cell number (from 7800 to 15600 cells ml^{-1}) might assimilate 1% of the initial nitrate supply, whereas the final doubling in cell number (from 500000 to 1000000 cells ml^{-1}) would utilize 50% (the *last* 50%) of the nitrate supply and would be followed by the stationary phase. The point is that if the culture is used to measure some physiological parameter of growing cells, the

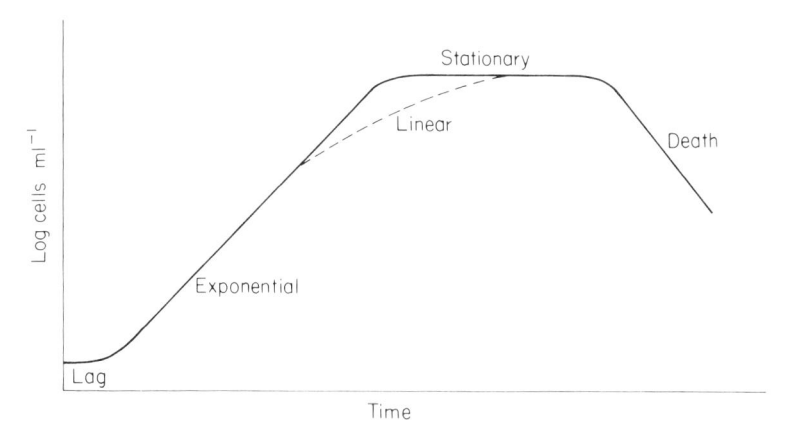

Fig. A2.2 Semi-log plot of a typical growth curve in a batch culture. (After Fogg 1965.)

investigator should remember that conditions are more nearly constant during early exponential growth and that cell physiology could exhibit some changes even before cessation of cell division signaled the onset of stationary phase.

The abrupt transition from exponential to stationary growth implied in the example above is the exception rather than the rule. In the usual case, the shift is more gradual as the growth rate gradually drops to zero (the linear phase of Fig. A2.2). In the case of high levels of inorganic nutrients, the cells could become so crowded that growth would be limited by the rate at which CO_2 diffused into the medium, or by the rate at which light reached cells shaded by other cells. Under these conditions growth would continue at reduced rates until some factor, perhaps nutrient depletion or the accumulation of toxic products, finally limited any further growth (Fogg 1965). In the case of nutrient depletion, the decrease in growth rate might be more abrupt and the period of declining growth rates shown in Fig. A2.3 (caused by low nutrient concentrations; see p. 34) would be very difficult to detect (O'Brien 1972). Batch cultures are extremely useful in autecological studies as long as the environment and possible cell-cycle effects (p. 155) are taken into account. They have been used most successfully to study the effect of parameters such as light intensity, temperature and salinity on growth rates

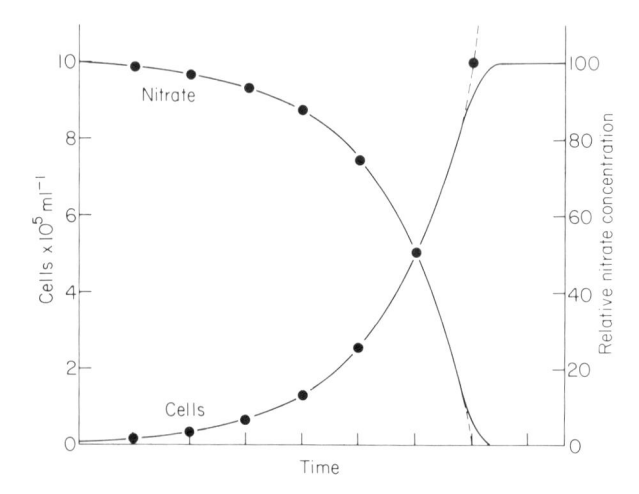

Fig. A2.3 Linear plot of the cell concentration and nitrate concentration in a batch culture ultimately limited by nitrate. (After O'Brien 1972.)

and to provide material for countless studies on ultrastructure, chemical composition and intermediary metabolism.

It is difficult to carry out laboratory, batch-culture experiments under the more 'natural' conditions of low cell and nutrient concentrations because there is insufficient cell material for many analyses. Large-scale or 'big-bag' culture systems have been developed to circumvent this problem and to provide systems in which interactions in natural planktonic communities can be studied. Large, translucent plastic bags are filled with 1–100 m^3 or more of water and suspended from floats in lakes or in the sea (Menzel & Case 1977). Changes in community structure, chemical composition and activity of the phytoplankton are monitored for several weeks and correlated with environmental parameters. Recently such systems have been used to test the effect of pollutants such as petroleum and copper on natural systems. Big-bag systems are cultures in the sense that a single, confined water mass is being monitored, but they improve on cultures in that natural communities are studied under 'natural' conditions. Wall effects are minimized because large volumes of water are involved.

Continuous cultures represent another approach to the problem of simulating natural conditions in a defined system. A chemostat is a continuous culture system in which fresh medium is constantly being added, replacing an equal volume of culture which is removed along with the cells it contains. The concentration of the nutrients in the fresh medium is suitably adjusted so one nutrient is present at potentially limiting concentrations. The dilution rate (ratio of flow rate to culture volume) is adjusted to be high enough to provide an adequate supply of nutrients, but not so high as to wash the cells out of the culture vessel. Under these conditions, the culture will reach a steady-state condition (cell concentration in the vessel does not change with time) in which the growth rate equals the dilution rate. For example, if the flow rate is 500 ml d^{-1} into a culture of 1000 ml, the dilution rate would be 0.5 d^{-1} and the specific growth rate, μ, would be 0.5 d^{-1} once the culture reached steady state. The specific growth rate can be expressed in terms of doubling time (DT) or generation time of the cells (DT = $\ln 2/\mu$ = 1.39 d) and also in terms of doublings or divisions per day (div. d^{-1} = $\mu/\ln 2$ = 0.72).

In most chemostats in a steady-state condition, the growth rate of the cells is limited by the low (steady-state) level of the limiting nutrient, which, in turn, is a function of its concentration in the fresh medium and of the dilution rate. The population density in a chemostat may be manipulated by changing the concentration of the limiting nutrient in the fresh medium. Suppose the concentration of the limiting nutrient were increased while the dilution rate remained constant: the increased nutrient supply would result in an increased growth rate of the cells which would increase the cell concentration in the chemostat. Soon, however, the higher cell density would increase the demand on the nutrient supply in the vessel to the point that the growth rate would return to its former value. Therefore, a new steady-state condition would be achieved at a higher cell density but at the same specific growth rate, since the dilution rate had not been changed.

The more usual situation is to manipulate the growth rate of the cells by changing the dilution rate. When the dilution rate is lowered, the supply of the limiting nutrient decreases and the specific growth rate decreases, eventually reaching a new steady-state condition in which it equals the new dilution rate. The cell density in the culture usually increases due to the decreased rate of cell removal and to a decrease in cell quota (p. 34). At infinitely low dilution rates, the chemostat becomes a batch culture in stationary phase (the growth rate drops to zero). If the dilution rate is increased from its initial value above, the increased flow of nutrients results in an increased specific growth rate which will equal the increased dilution rate once a steady state is achieved. Cell density decreases due to higher rates of cell removal and to an increase in cell quota (p. 34). Further increases in dilution rate result in higher growth rates until the cells are growing at their nutrient-saturated rates for the given conditions (light intensity, temperature, etc.). Any further increase in dilution rate will result in the wash-out of the cells because they cannot grow fast enough to keep up with the dilution.

The tremendous advantage of the chemostat over batch cultures is that one can study the physiology of cells actually growing under steady-state, nutrient-limited conditions. Thus, chemostats simulate natural conditions much more closely than batch cultures in which nutrient supply is either very high (early exponential growth) or rapidly dropping to negligible levels (late exponential growth). With chemostat cultures one can select any number of different growth rates representing an equal number of nutrient-limited states simply by adjusting the dilution rate and waiting for the culture to reach a steady state. In batch cultures the cells shift from nutrient-saturated growth to stationary phase (nutrient starved) often within a single cell generation. As a result it is difficult to study the intermediate cases of slow growth under nutrient-limited conditions with batch cultures. Batch cultures, however, are the preferred system for studying cells under nutrient-saturated conditions because chemostats are on the verge of washing out once growth rates approach nutrient-saturated rates.

Although chemostats are a major improvement over batch cultures for some studies, it is an oversimplification to suggest that their steady-state condition simulates natural conditions in which there are temporal variations in many parameters, especially in a 24 h time frame. Chemostats which are grown under light–dark cycles (cyclostats) will more closely simulate natural conditions than those exposed to continuous light. The light–dark cycle serves to phase (partially synchronize) cell division in the culture so that cell density in the culture fluctuates over the 24 h cycle; cells increase in the culture during the period of cell division and then decrease as they are washed out during the rest of the cycle. Other parameters such as nutrient concentrations also may fluctuate during the daily cycle, although the system is still in steady state when considered on a time-scale of days rather than hours. Qualitatively the cyclostat functions much like the steady-state chemostat with the additional proviso that growth may be limited by the dark period, in addition to (or instead of) nutrient levels. Quantitative treatment of cyclostat theory is discussed by Chisholm et al. (1975).

Dialysis culture, yet another system which has been introduced recently, combines several advantages of other culture systems. A small water sample (100 ml to several litres) is placed in a container consisting either in part or totally of a semi-permeable membrane; dialysis tubing is often used. The water sample may consist of a natural phytoplankton assemblage or may contain a dilute suspension of a unialgal or axenic culture. In either case, the sample is usually suspended in natural water and monitored for several days. During incubation, nutrients, dissolved gases and small organic molecules diffuse across the membrane to support growth under close to natural conditions. The system has the advantage of a small, inexpensive and easily handled culture apparatus, a continuous supply of nutrients at natural concentrations, no accumulation of toxic waste products which could inhibit growth, and a confined population or community of phytoplankton which can be monitored. Although high cell concentrations can be achieved in dialysis cultures, for most purposes it is best to confine observations to the early growth phases even though limited amounts of cell material will preclude some measurements. Dialysis cultures can also be used as a bioassay system to monitor water quality, as an exclosure system to monitor phytoplankton growth in the absence of grazing zooplankton and as a system to study species interactions involving diffusible inhibitory or stimulatory substances (Jensen et al. 1972).

REFERENCES

CHISHOLM S.W., STROSS R.G. & NOBBS P.A. (1975) Light/dark-phased cell division in *Euglena gracilis* (Z) (Euglenophyceae) in PO_4-limited continuous culture. *J. Phycol.*, **11**, 367–73.

FOGG G.E. (1965) *Algal Cultures and Phytoplankton Ecology*. University of Wisconsin Press, Madison.

HELLEBUST J.A. & CRAIGIE J.S. (eds) (1978) *Handbook of Phycological Methods. Physiological and Biochemical Methods*. Cambridge University Press, Cambridge.

JEFFREY S.W. (1981) An improved thin-layer chromatographic technique for marine phytoplankton pigments. *Limnol. Oceanogr.*, **26**, 191–7.

JENSEN A., RYSTAD B. & SKOGLUND L. (1972) The use of dialysis culture in phytoplankton studies. *J. exp. mar. Biol. Ecol.*, **8**, 241–8.

LEAN D.R.S. & BURNISON B.K. (1979) An evaluation of errors in the ^{14}C method of primary production measurement. *Limnol. Oceanogr.*, **24**, 917–28.

MENZEL D.W. & CASE J. (1977) Concept and design: Controlled ecosystem pollution experiment. *Bull. Mar. Sci.*, **27**, 1–7.

O'BRIEN W.J. (1972) Limiting factors in phytoplankton algae: Their meaning and measurement. *Science*, **178**, 616–17.

ODUM E.P. (1971) *Fundamentals of Ecology*, 3rd edn. W.B. Saunders Co., Philadelphia.
STEIN J.R. (ed.) (1973) *Handbook of Phycological Methods. Culture Methods and Growth Measurements*. Cambridge University Press, Cambridge.
STRICKLAND J.D.H. & PARSONS T.R. (1972) *A Practical Handbook of Sea-water Analysis*. Bull. fish. res. Bd. Can., 167.
THEODÓRSSON P. & BJARNASON J.O. (1975) The acid-bubbling method for primary productivity measurements modified and tested. *Limnol. Oceanogr.*, **20**, 1018–19.
VOLLENWEIDER R.A. (ed.) (1969) *A Manual on Methods for Measuring Primary Production in Aquatic Environments*. Int. Biol. Programme Handbook 12. Blackwell Scientific Publications, Oxford.
WHITNEY D.E. & DARLEY W.M. (1979) A method for the determination of chlorophyll *a* in samples containing degradation products. *Limnol. Oceanogr.*, **24**, 183–6.

Index

Achnanthes 145
Adaptation see Complementary chromatic adaptation; Mercury; Nutrients, limitation; Sun-shade adaptation
Agar 14, 98, 149
Agarum 107
Akinetes 5, 63, 68, 134
Alginic acid 12, 98, 149
Alkaline phosphatase 40–1, 42, 80–1
Allelopathy 51, 59–60, 72
Alternation of generations 3–4, 11, 13, 88, 101–2, 107, 147
Ammonium
  light and uptake 96, 118
  and nitrogen fixation 64, 126, 128–30
  preferred nitrogen source 42, 95
  uptake by corals 118
  uptake kinetics 33, 37–8, 95–6, 147
Amphora 145
Anabaena 22, 64, 66, 67, 144
Anabaena azollae 126–8, 130
Anthoceros 128–9
Aphanizomenon 22, 64, 66, 67, 144
Ascophyllum 11
Assimilation number 31, 38, 80–1, 122
Asterionella 11, 22, 69
ATP 40–1, 157
Aufwuchs see Periphyton
Azolla 126–9

Bacillariophyceae 2, 10–1, 72, 154
  habitats 21–2, 49, 71, 74–5, 77, 78, 82–5, 106, 132, 134, 140, 145, 149
  physiology 29, 45, 49, 55, 59, 85
  silicon 43, 55–6, 57
Batrachospermum 78
Bioluminescence 9, 55–6
Biomass, measuring 76, 82–3, 88, 90, 133, 155–7
Blasia 128–9
Blooms 24, 60, 67, 71, 144; see also Red tides
Blue-green algae see Cyanophyceae
Botryococcus 22, 58
Brown algae see Phaeophyceae
Buoyancy see Suspension

Calcification 12, 13, 118–9, 159
Caloglossa 100
Calothrix 75, 140–1
Carbohydrate 31, 37, 44
Carbon
  cell quota 36–8, 43–4
  inorganic 48, 67, 99, 149, 157–8
Carotenoids 8, 27, 29, 93, 119, 135, 157
Carposporophyte 13
Carrageenan 14, 98, 149
Caulerpa 89
Cephalodia 120
Ceramium 100
Ceratium 22
Chaetoceros 22, 55
Chara 5, 85–6
Characium 133
Charales 1, 6, 85–6
Charophyceae 2, 5–7, 153
Chelators 44, 51
Chlainomonas 135
Chlamydomonas 6, 22, 75, 133, 135–6
Chlorella 111–5, 140, 148–9
Chlorococcum 133
Chloromonas 135–6
Chlorophyceae 2, 3, 5–7, 9, 153
  habitats 13, 21–2, 67, 71, 74–5, 89, 119, 120, 132, 135
  physiology 45, 93, 95
Chlorophyll see also Assimilation number; Biomass, measuring; Sun-shade adaptation
  diel periodicity in synthesis 54–5
  measurement 76, 133, 156–7
  nutrient-limitation 37–8, 43–4
  temperature 46
Chloroplasts, symbiotic in sea slugs 119
Chondrus 89, 94, 106–7, 150
Chroomonas 54
Chrysophyceae 2, 9–10, 12, 153
  habitats 21–2, 74, 135
  physiology 43, 45
Circadian rhythms see Periodicity
Cladonia 121
Cladophora 6, 75, 77, 78, 145
Closterium 75
Coccolithophorids see Prymnesiophyceae
Coccomyxa 130

164

Cocconeis 75
Codium 6, 89, 94, 97, 102
Colacium 7
Coleochaete 6–7, 75
Compensation depth 27, 93
Compensation light intensity (I_c) 27, 28, 115
Competition 57, 68–70, 74, 86, 105–8, 141
Complementary chromatic adaptation 31–2, 93
Convergence zones 25, 61–2
Convoluta 111–4
Copper 144–5, 146
Corals 111, 115–9
Cosmarium 22
Cryptomonas 22, 54
Cryptophyceae 13, 21–2, 32, 135, 154
Cultures
   batch 34, 159–62
   big-bag enclosures 41, 146, 161
   chemostat 34–8, 39, 40, 46–7, 146–7, 161
   compared to natural systems 19–20, 22–3, 56–7, 59, 60, 76, 82–3, 88–9, 146, 148, 155, 161–2
   cyclostat 57, 162
   dialysis 60, 79, 162
   mass 147–50
   synchronized 43, 44, 53, 57, 155, 162
Currents 74, 78, 79, 94, 102–3
Cyanidium 137, 141
Cyanophyceae 4–5, 153
   habitats 21–2, 71, 74–5, 82, 132–4, 136–40
   physiology 32, 45, 60, 63–67, 78
   symbiotic 120, 125, 126–30
Cyanophycin 63, 126
Cyclotella 22, 69

DCMU 113, 118, 139
Desiccation
   microalgae 78, 134–5, 140
   seaweeds 96, 97, 98–100, 107
   symbiotic algae 122–6, 129
Desmids 6, 21, 82
Diatoms *see* Bacillariophyceae
Dinobryon 10, 22
Dinoflagellates *see* Pyrrhophyceae
Ditylum 55, 58
Divergence zones 25
Dunaliella 36–7, 49, 148

Ectocarpus 11, 144–5
Egregia 108
Enteromorpha 10, 89, 106, 144, 146
Epilimnion 23–4, 66
Epipelic algae 76, 82–5
Epiphytic algae *see* Periphyton
Epipsammic algae 82
Euglena 8, 22, 75
Euglenophyceae 2, 3, 7–8, 13, 153
   habitats 22, 67, 74–5, 82, 132
   physiology 42, 45

Euphotic zone 23, 25, 27, 50, 58–9, 66, 67, 71
Eustigmatophyceae 12, 154
Eutrophic waters 39–40, 44, 66–7, 72, 80, 85, 145

Floridoside 98
Flotation *see* Suspension
Fouling 144–5
Fragilaria 22
Frustule 10, 43
Fucoidan 98
Fucoxanthin *see* Carotenoids
Fucus 11, 89, 94, 96, 97, 98–9, 106–7, 146

Gas bladders 11, 88
Gas vacuoles (vesicles) 65–7
Gelidium 89
Geotaxis 82
Gigartina 89, 150
Gloeotrichia 67
Glutamate dehydrogenase 37–8
Glutamine 64, 113–4, 126, 130
Glutamine synthetase 64, 126, 130
Glycerol 49, 112, 116–7, 148
Glycerophosphate 42
Gomphonema 75
Gonyaulax 9, 60, 61, 143
Gracilaria 96
Grazing 16
   on microalgae 54, 56, 58, 59, 70, 141, 148, 157, 162
   on seaweeds 106–8, 146, 149–50
Green algae *see* Charophyceae; Chlorophyceae
Growth
   curve 159–61
   rate 16
      Chlorella in symbiosis 113–5
      calculate 53, 161
      lichens 122–3, 125
      light effects 31, 46, 101
      Nostoc–bryophyte symbiosis 129, 130
      nutrient-limited 34–44, 69, 147, 160–2
      nutrient-saturated 35, 162
      r- and K-selected species 72
      seaweeds 88, 91–2, 96–7, 100, 103, 105
      specific growth rate (μ) 34, 161
      temperature effects 45–8, 96–7, 138
      versus yield 148
Gymnodinium 60, 143–4

Halimeda 6
Hantzschia 82–3
Haptera 88, 102
Harveyella 81
Heterocyst 64, 125–6, 127, 129–30
Heterotrophy 7, 8, 9, 49–50, 68, 85, 135, 141
Hot springs 136–41
Hydra 111–4

165

Hypnozygote 61
Hypolimnion 23–4, 63

Iron 35, 44

Janczewskia 81

Kelps 11, 88, 90–1, 101–2, 104
Klebsormidium 6–7
K-selection 72

Laminaran 91–2
Laminaria 89, 91–2, 99, 103, 105, 108
Langmuir circulations 25, 30, 58–9, 67
Laurencia 81
Lichens 119–26
Light
  intensity 65, 78, 102, 135; see also
    Photoinhibition; Sun-shade adaptation
  attenuation 25, 84, 136
  growth 31, 46, 101
  limiting factor 23, 62, 74, 77–8, 84, 86,
    101–2, 105, 133, 148
  metabolism 31
  nitrogen fixation 64, 126, 129, 140
  nutrient uptake 33, 55, 96, 118
  photosynthesis 26–31, 92–4, 115–6,
    138–9
  temperature 46
  quality 25, 29, 31–2, 93, 101, 102
Limiting factors 15–9
Lipid 37, 44, 58, 116–7, 134, 137
Lithodesmium 55
Lyngbya 144

Macrocystis 11, 89, 95, 104, 149
Mallomonas 54
Mannitol 49, 91–2, 98, 104, 114, 121–2, 124
Mastigocladus 140
Mercury 147
Microcystis 66, 67, 144

Navicula 75, 145
Nereocystis 107–8
Nitella 75, 85–6
Nitrate
  and Azolla 128
  reduction 42
  seaweeds 91–2, 95–6
  uptake kinetics 32–3, 37–8, 95–6
Nitrate reductase 42
Nitrite 42
Nitrogen 42
  Azolla as fertilizer 128
  cell quota 35–8, 43
  fixation 63–5, 85, 125–30, 132, 140
    inhibition by nitrate 64, 126, 128
    light dependence 64, 126, 129, 140

and hydrocolloid production 150
limitation 32, 35–8, 40, 42, 68, 84, 91–2,
  96, 103, 126, 130, 147
specific uptake rate 37–8
Nitrogenase 63–4, 126, 128, 130
Nitzschia 75
Nori 14, 102
Nostoc 120, 125, 128–9
Nutrients see also Currents; Nitrogen;
  Phosphorus; Silicon; Suspension
and assimilation number 31
batch cultures 159–60
cell quota 34–8, 43–4, 46–8, 161
and competitive interactions 68–70
enrichments 40–2, 44, 65–6, 78, 84, 91,
  125
and epiphytic associations 81
lichens 125
light and uptake 33, 55, 96, 118
limitation 24, 33–44, 46–7, 71, 91–2, 96,
  126, 150, 159–62
luxury consumption 42
and photoinhibition 27
saturated growth 34, 40, 43–4, 47, 162
and seasonal succession 71
standing stock 18, 20, 23, 33, 39, 74,
  91–2, 125
starvation 34, 43–4, 162
symbiosis 118, 125
and temperature 46–7
uptake kinetics 32–4, 37–9, 43, 55–6, 57,
  69, 85, 95–6, 147

Ochromonas 9, 54
Odonthalia 81
Oedogonium 6, 75, 77, 78
Oligotrophic waters 33, 39–40, 44, 72, 80,
  157, 158
Oocystis 54
Organic excretion 45, 51, 60, 117, 159
Organic exudation 104–5
Oscillatoria 63, 64, 65, 66–7, 75, 141
Osmoregulation 49, 98

Paradox of the plankton 68–70
Paramecium 111–5
Parasites 81
Pedinomonas 7
Peltigera 124–6, 130
Pelvetia 99
Peridinium 22, 60
Periodicity
  circadian rhythms 9, 56–7, 82, 94
  diel 31, 53–8, 82–3, 84
  and paradox of plankton 70
Periphyton 77–82
  artificial substrates 78–9, 80–1, 147
  epilithic algae 77–9
  epiphytic algae 79–82, 89, 149
Periwinkles 106–7
pH 48, 129, 132–3, 137, 141, 158
Phaeodactylum 148

Phaeophyceae 2, 3, 11–2, 154
  habitat 13, 89
  physiology 93, 94, 95, 98, 104
Phosphorus 42; see also Alkaline
    phosphatase
  cell quota 35, 37, 43
  and charophytes 86
  debt 40–1
  light and uptake 118
  limitation 32, 35–6, 38, 40–1, 69, 78, 85
  luxury consumption 42
  polyphosphate 42
  specific uptake rate 38–9
  translocation in kelps 95
  uptake by corals 118
  uptake kinetics 33, 37–8, 78, 85
Photoinhibition 27, 29, 84, 92, 136, 138–9
Photoperiod 28, 31, 97, 102
Photoprotection 27, 135
Photosynthesis see also Light; Respiration;
    Sun-shade adaptation
  anoxygenic 63, 137
  capacity (P_{max}) 26–31, 37–8, 41, 55,
    92–4, 115–6
  and desiccation 99, 124
  versus light intensity 26–31, 84, 91–4,
    115–6, 139–40
  measurement 41, 76, 84, 88, 90–1, 99,
    157–9
  net and gross 27, 157–9
  nutrient-limitation effects 37–8, 96, 130
  quotient 158
  and temperature 97, 138–9
Phototaxis 54, 62, 82, 113, 136
Phycobiliproteins 2, 5, 13, 141
  and light 32, 93, 139
  and nitrogen 63, 64, 96, 129
Phytochrome 102
Phytoplankton 21, 22–3, 70, 78
  net/nanoplankton 21
Pigmentation see Carotenoids; Chlorophyll;
    Complementary chromatic adaptation;
    Phycobiliproteins; Sun-shade
    adaptation
Platymonas 43, 49, 111–4
Plectonema 139–40
Pleurocapsa 141
Pollution 125, 145–7, 148, 149, 161, 162
Polyols 49, 121–4
Porphyra 14, 89, 94, 98, 102, 147
Postelsia 89
Potassium 48–9, 58, 65, 67, 98, 149
Prasinophyte 7
Prorocentrum 60, 61–2
Protein 31, 37, 119, 137, 149
Prototheca 144
Prymnesiophyceae 2, 12, 21, 59, 154, 159
Prymnesium 144
Pyrocystis 58
Pyrodinium 60
Pyrrhophyceae 2, 8–9, 72, 153; see also
    Bioluminescence; Red tides; Toxins
  habitats 21–2, 71, 111, 115–9, 135
  physiology 45

Raphe 11
Red algae see Rhodophyceae
Redfield ratio 40
Red tides 60–2, 143–4
Respiration 27–8, 37, 51, 62, 71, 90, 122–3,
    125, 157–9
Resting cells 17, 67–8, 134–5; see also
    Akinetes; Hypnozygotes
Rhodophyceae 4, 13–4, 81, 89, 93, 95, 96,
    97, 98, 154
Rhythms see Periodicity
Ribulose bisphosphate carboxylase 28–9,
    64, 119
r-selection 72, 107

Salinity 48–9, 61–2, 85, 97–8
Sargassum 11, 89
Saxitoxin 143–4
Scales 10, 12, 43
Scenedesmus 22, 35–6, 148
Sea anemones 111–3, 115–9
Sea urchins 106–8, 149
Seaweeds 88–108, 149–50
Silicic acid see Silicon
Silicon 43–4
  cell quota 43–4
  limitation 32, 43–4, 69
  and seasonal succession 71
  and suspension 59
  uptake kinetics 33, 43, 55–6, 57
Skeletonema 22, 43–4, 46, 55, 147, 148
Sodium, requirement 48
Soredium 120–1
Spirogyra 6, 82
Spirulina 75, 148
Statistical analysis 19
Statospore 9, 68
Stigeoclonium 75, 77, 78
Streams 77–9
Succession 60, 70–1, 78, 105–8, 132–3
Sun-shade adaptation
  corals 115–6
  hot-springs algae 139–40
  kelp gametophytes 101–2
  phytoplankton 27–9, 31
  Prorocentrum 62
  sea anemones 113
  seaweeds 93–4
Suspension 54, 58–9, 65–7, 70
Symbiosis 2, 8, 111–30
Synechococcus 138–9, 140, 141
Synura 10, 22, 43, 145

Tabellaria 75
Temperature 45–8, 96–7; see also
    Thermocline
  and assimilation number 31
  hot-springs algae 136–40
  hypnozygote germination 61
  lichens 124–5
  and light 46, 97
  in mass cultures 148
  and nutrients 33, 46–8

167

phytoplankton   45–8
seaweeds   91, 96–7, 101
snow algae   135–6
stream algae   78
and water-column stability   23–4, 71
Tetracystis   133
Tetrasporophyte   13
Thalassiosira   38, 55
Thermocline   23–4, 65–6
Thermophiles   137–41
Tolerance limits   16, 97, 98–9, 100, 137
Toxins   9, 60–1, 67, 143–4
Trace elements   44
Translocation
  in kelps   11, 91, 95, 104
  in symbiosis   81–2, 112, 114, 116–7, 121–4, 129
Trebouxia   120, 121
Tribonema   78
Tridacna   111, 117
Tychoplankton   74

Ulothrix   75, 77, 144
Ulva   6, 89, 94, 146

Upwelling   23
Urea   42
Uric acid   113–4

Vaucheria   9
Vertical migration   54
  epipelic diatoms   82–3, 84
  Prorocentrum   62
  snow algae   136
Vitamins   8, 9, 35, 45

Xanthophyceae   2, 9, 12, 74, 132, 153

Zoochlorella *see Chlorella*
Zooplankton   21, 58, 70; *see also* Grazing
Zooxanthella   111–2, 115–9
Zooxanthellae *see Zooxanthella*
Zygospore   68, 136